ENVIRONMENTAL BIOTECHNOLOGY

ENVIRONMENTAL BIOTECHNOLOGY

ENVIRONMENTAL BIOTECHNOLOGY

Editor
Rajmohan Joshi

ISHA

ISHA BOOKS
DELHI-110033

Environmental Biotechnology

ISBN: 81-8205-382-X

Rs. 690

Published in 2006 in India by
Isha Books
D-43, Prithviraj Road, Adarsh Nagar, Delhi-110033
E-mail : ishabooks@hotmail.com

Laser Typesetting at: Master Computers, Delhi
Printed at: Salasar Imaging System, Delhi

Preface

Biotechnology is a research oriented science, a combination of biology and technology. There are many application of biotechnology such as developing various medicines, vaccines and diagnostics, increasing productivity, improving energy production and conservation. Environmental biotechnology is the application of biotechnology processes or products to any aspects of the environment. It is the development, use and regulation of biological systems for remediation of contaminated environments and for environment-friendly processes.

The major environmental issues which biotechnology is being utilised to tackle include: tackling pollution and contamination which has already affected land, water and air; processing of current waste streams from industry, agricultural and domestic homes; devising new reduced-waste, renewable, environmentally friendly technologies to replace current industrial processes and the improvement of crops or livestock through genetic manipulation. Much of the biotechnology used within the environmental sector exploits the metabolism of micro-organisms. A great deal of the reactions are catabolic whereby organic molecules are broken down to smaller molecules as part of waste treatment, pollution control or land decontamination.

This book provides deeper insight into the concepts and applications of environmental biotechnology. Designed for courses at undergraduate and graduate levels, this book will also serve as an essential reference for environmental microbiologists, environmental engineers, as well as those interested in water and wastewater treatment, and biotechnology.

Editor

Contents

1

Introduction to Ecology and Ecosystem

Ecology is the scientific study of the interactions between organisms and their environments. The scientific nature of ecology involves using observations and experiments to test hypothetical explanations of ecological phenomena.

Ecology has a long history of being a descriptive science but is young as an experimental science. Some scientists devise mathematical models that include important variables and hypothetical relationships; usually studied with the aid of a computer. Mathematical models are also used to simulate large-scale experiments that are impossible to conduct in the field; however, the basic information on which the models are based still must be obtained through fieldwork.

Divisions of Ecology

Ecology can be divided into four increasingly comprehensive levels of inquiry:

- Organismal ecology studies the behavioral, physiological, and morphological ways individuals meet abiotic environmental challenges.
  - The distribution of organisms is limited by their tolerance of abiotic conditions.

- Population ecology studies groups of individuals of the same species living in a particular geographic area.
  - Questions in population ecology concern factors that affect population size and composition
- Community ecology studies all organisms that inhabit a particular area.
  - Questions concern predation, competition, disease, and other ways in which interactions among organisms affect community structure and organization
- Ecosystem ecology studies all abiotic factors as well as communities of organisms in an area.
  - Questions concern energy flow and chemical cycling among the abiotic and biotic components

Ecological study is multidisciplinary in nature, encompassing genetics, evolution, physiology, behavior, chemistry, physics, geology, and meteorology. Ecology provides a scientific context for evaluating environmental issues. Although distinct, basic ecology and environmental issues have many connections.

- To properly address environmental problems, it is necessary to understand the relationships between organisms and their environments.
- The realization that Earth is a finite resource and that pollution, overpopulation, and habitat destruction are threatening that resource is a concern of most people.

The environment of an organism includes both biotic and abiotic components. Biotic components include all other organisms that are a part of any individual organism's environment. Abiotic components are the nonliving chemical and physical factors (e.g., temperature, light, water, nutrients) to which an organism is exposed.

The interactions between organisms and their environments include how the environment affects an organism and how an organism can change the environment. Photosynthetic bacteria began to use sunlight for energy about three billion years ago. Oxygen, a by-product of photosynthesis, accumulated and resulted in an aerobic atmosphere. The shading of the forest floor by trees sometimes makes the floor unsuitable (due to reduced light) for their offspring to grow.

Short-term (ecological time) interactions of organisms with their environments could have long-term (evolutionary time) effects through natural selection. For example, predator-prey interactions may affect gene pools where individuals with protective coloration would become more prevalent. The current distribution and abundance of organisms are products of long-term evolutionary changes and ongoing interactions.

Abiotic Factors of the Biosphere

The biosphere is the global ecosystem-the sum of all Earth's ecosystems. It is a thin layer consisting of the atmosphere to an altitude of a few kilometers; the land down to and including water-bearing rocks at least 1500 meters below ground, lakes and streams, caves; and the oceans to a depth of several kilometers.

Climate and other abiotic factors are important determinants of the biosphere's distribution of organisms. Global and regional patterns reflect differences in climate and other abiotic factors.

— Abiotic factors such as temperature, humidity, salinity, and light influence the distribution of organisms.

— The patchiness of the global biosphere illustrates how different physical environments produce a mosaic of habitats.

Some of the important abiotic factors that affect distribution of species include: temperature, water, sunlight, wind, rocks and soil, and periodic disturbances.

Temperature

Environmental temperature affects biological processes and body temperature.

— Most organisms are unable to regulate their body temperature precisely

— Temperature greatly affects metabolism: few organisms have active metabolisms at temperatures close to 0°C, and temperatures above 45°C denature most essential enzymes.

— The actual body temperature of ectotherms is affected by heat exchange with the environment.

— Most animals maintain a body temperature only a few degrees above or below ambient temperature.

— Even endotherms function best within the environmental temperature range to which they are adapted.

Water

Water is essential for life and adaptations for water balance and conservation help determine a species' habitat range. Marine and freshwater animals face the problems of regulating intracellular osmolarity; terrestrial animals face the problem of desiccation. Water selectively reflects and absorbs certain wavelengths; therefore, most photosynthesis occurs near the water surface.

Sunlight

Sunlight provides the energy that drives nearly all ecosystems although only photosynthetic organisms use it directly as an energy source. Light is not the most important limiting factor in terrestrial environments but may play a role as in the reduction of competition due to shading in a forest. In aquatic environments, the distribution of photosynthetic organisms is limited by the intensity and quality of light. The physiology, development, and behavior of many animals and plants are often sensitive to photoperiod.

Wind

Wind amplifies the effects of temperature by increasing heat loss by evaporation and convection. Wind also increases the evaporation rate of animals and transpiration rate of plants, resulting in more rapid water loss. Mechanical pressure of wind can affect plant morphology (for example, inhibiting growth of limbs on windward side of trees).

Rocks and soil

The physical structure, pH, and mineral composition of soil limit distribution of plants and hence animals that feed on those plants.

— The composition of the substrate in a stream or river greatly influences the water chemistry, which in turn influences the plants and animals.

— The type of substrate influences what animals can attach or burrow in intertidal zones.

Periodic disturbances: Catastrophic disturbances such as fire, hurricanes, typhoons, and volcanic eruptions can devastate biological communities.

— After the disturbance, the area is recolonized by organisms or repopulated by survivors., but the structure of the community undergoes a succession of changes.

— Those disturbances that are infrequent (volcanic eruptions) do not elicit adaptations. Adaptations do evolve to periodically recurring disturbances such as fires.

Climate and organisms distribution

Climate is the prevailing weather conditions at a locality. The major components of climate are temperature, water, light, and wind. Climate has a major impact on the distribution of organisms. A climatograph plots temperature and rainfall in a particular region.

— Usually plotted in terms of annual means

— Must be careful to distinguish a correlation between climate variables and biomes from causation.

— Biomes are the major types of ecosystems typical of broad geographic areas.

— Only a detailed study of a species' tolerances to water and temperature ranges could establish the direct effects of these variables.

— The fact that biomes overlap on climatographs indicate that factors other than mean temperature and rainfall play roles in determining biome location (e.g., soil composition).

Global climate patterns: Solar energy input and the Earth's movement in space determine the planet's global

climate patterns. About 50% of the solar energy that reaches the atmosphere's upper layers is absorbed before it reaches the surface.

— Ultraviolet and certain other wavelengths are more readily absorbed by oxygen and ozone than other wavelengths.

— Some of the solar energy that reaches the Earth's surface is reflected back into the atmosphere; large amounts are absorbed by land, water, and organisms.

The atmosphere, land, and water are heated when they absorb solar energy. This heating establishes the temperature variations, air movement cycles, and evaporation of water responsible for the latitudinal variations in climate.

— Latitudinal variation in the intensity of sunlight results from the Earth's spherical shape; seasonal variation in solar radiation in the Northern and Southern Hemispheres are due to the Earth's tilt of 23.5° relative to its plane of orbit. The tilt causes solar radiation to change daily as the Earth rotates around the sun.

— Only the tropics (23.5°N to 23.5°S) receive sunlight from directly overhead year round. The tropics receive the greatest annual input of solar radiation and show the least seasonal variation; only small variations in daylength and temperature occur.

— Seasonal variation in light and temperature increases steadily toward the poles; polar regions have long, cold winters with periods of continual darkness and short summers with periods of continual light. A global circulation of air which creates precipitation and winds results from the intense solar radiation near the equator

— Evaporation of surface water due to high tropical temperatures causes warm, wet air masses to rise near the equator; this rising air creates an area of light, shifting winds (doldrums) along the equator.

— As these warm air masses rise, they expand and cool; cool air can hold less water vapor so the rising air masses drop large amounts of rain in the tropics.

— The cool, dry air masses flow toward the poles at high altitudes; they continue to cool as they move farther from the equator.

— Air mass density increases as they become cooler; they begin to descend toward the surface as cool, dry air masses at about 30° latitude.

- The air masses absorb water as they descend, thus creating arid climates around 30°N and 30°S.
- Some of the descending air masses flows toward the poles at low altitudes, the rest flows toward the equator.
- The air masses flowing toward the poles are warmed and rise again at about 60°N and 60°S; as these masses of air begin to cool with increased altitude, the water vapor is lost as precipitation.
- As air from this third cell reaches the higher altitude and cools, it flows toward the poles where it descends and flows back toward the equator.

Local and seasonal effects on climate: Although global climate patterns explain the geographic distribution of major biomes, local variations due to bodies of water and topographical features create a regional patchiness in climatic conditions.

- The regional and local variations in climate and soil have a major influence on less widely distributed communities and individual species. Proximity of large bodies of water affect local climates.
- Ocean currents are generated by the Earth's rotation and may heat or cool (depending on whether they are tropical or polar currents) air masses passing over them toward land; evaporation is also greater over the ocean than over land.
- Coastal areas are more moist than inland areas at the same latitude; the California current flows from southward along the western U.S. and helps form a cool, moist climate in this area.
- During warm summer days, air over land heats faster than that over the ocean or large inland lakes; this warmer air rises, drawing a cool breeze from the water across the land.

Topographical variations, such as mountains, also exert an influence on solar radiation, local temperature and rainfall.

— In the Northern Hemisphere, south-facing slopes receive more sunlight and are therefore warmer and drier than north-facing slopes.

- — The vegetation differs with south-facing slopes being covered by shrubby, drought resistant plants while the north-facing slopes have forests.

— Air temperature declines 6° C for each 1000m increase in elevation.

- — This parallels the decline in temperature associated with increasing latitude.
- — For this reason, mountain communities are similar to those at lower elevations farther from the equator.

— When warm, moist air moves over a mountain, its altitude is increased and it cools; cooling causes the air to release its moisture as rain on the windward side.

— The cooler, drier air then flows down the leeward side of the mountain; the cool dry air is warmed and absorbs moisture, producing the rainshadow.

— Deserts commonly occur on leeward sides of mountains.

The Earth's orbit around the sun causes seasonal changes in local conditions.

— The changing angle of the sun causes slight shifts in the wet and dry air masses on either side of the equator.

- — These shifts result in the wet and dry seasons at 20° latitude where tropical deciduous forests grow.

— Seasonal changes in wind patterns produce variations in ocean currents, sometimes causing upwellings that bring nutrient-rich, cold water from the deep ocean layer to the surface.

— Seasonal temperature changes also cause the temperature profiles that develop during the summer in temperate zone ponds and lakes.

— These profiles reverse in the autumn and spring resulting in biannual mixing that brings nutrient-rich water from the bottom to the top

Climate also varies on a smaller scale, the microclimate. Microclimate refers to small areas within a habitat that may have very different conditions than the overall area (e.g., under a rock, a forest floor).

— Cleared areas in a forest generally show greater temperature extremes than the shaded forest floor due to greater solar radiation and wind currents.

— Low-lying areas are usually moister than high ground and support different forms of vegetation.

— The area under a large stone or log is protected from extremes of temperature and moisture; a large number of small organisms usually live in such sheltered areas.

Worldwide Distribution of Aquatic and Terrestrial Biomes

Aquatic biomes occupy the largest part of the biosphere Life arose in water and evolved there for almost three billion years before moving into terrestrial habitats. Aquatic biomes still account for the largest part of the biosphere. Freshwater and marine biomes are distinguished on the basis of physical and chemical differences.

— Freshwater biomes have a salt concentration less than 1%; marine biomes average 3% salt concentration.

The oceans cover 75% of Earth's surface and contain the marine biomes.

— Evaporation from the oceans provides most of the rainfall.

— Ocean temperatures greatly effect world climate and wind patterns.

— Marine algae ad photosynthetic bacteria consume large amounts of the atmospheric carbon dioxide and produce a major portion of the world's oxygen. Freshwater biomes are closely linked to the soils and biotic components of the terrestrial biomes in which they are located or through which they flow.

— Runoff from terrestrial habitats creates streams and rivers.

— Accumulated runoff creates ponds and lakes.

— Freshwater biomes are also influenced by the patterns and speed of water flow and the climate to which they are exposed.

Stratification of aquatic biomes

Aquatic biomes often exhibit pronounced vertical stratification.

— There is a decrease in light intensity with increasing depth as light is absorbed by the water and suspended microorganisms. This divides bodies of water into two layers:

  — The photic zone is the upper layer where light is sufficient for photosynthesis.

  — The lower aphotic zone receives little light and no photosynthesis occurs.

— Water temperature tends to be stratified, especially during winter and summer.

— Heat energy from sunlight warms the upper layers of water as far as it penetrates; the deeper waters remain cold.

- The thermocline is a narrow vertical zone between the warmer and colder waters where a rapid temperature change occurs.
- At the bottom of all aquatic biomes is the benthic zone, which is occupied by communities of benthos organisms that consume dead organic matter.

Freshwater biomes

Standing bodies of water vary greatly in size, from ponds to large lakes. Ponds and lakes usually exhibit a significant vertical stratification in light penetration and water temperature. The distribution of plants and animals within a pond or lake also shows a stratification based on water depth and distance from the shore.

- The littoral zone is shallow, well-lit waters close to shore.
    - Characterized by the presence of rooted and floating vegetation, and a diverse attached algal community (especially diatoms)
    - There is a diverse animal fauna including suspension feeders (clams); herbivorous grazers (snails); and herbivorous and carnivorous insects, crustaceans, fishes, and amphibians.
    - Some reptiles, water fowl, and mammals also frequent this zone.
- The limnetic zone is the open, well-lit waters away from shore.
    - Occupants include photosynthetic phytoplankton (algae and cyanobacteria), zooplankton (rotifers and small crustaceans) that graze on phytoplankton, and small fish which feed on the zooplankton.
    - Occasional visitors to this zone are large fish, turtles, snakes, and piscivorous birds.

— The profundal zone is the deep, aphotic zone lying beneath the limnetic zone.

— This is an area of decomposition where detritus is broken down.

— Water temperature is usually cold and oxygen is low due to cellular respiration of decomposers.

— Mineral nutrients are usually plentiful due to decomposition of detritus.

— Waters of the profundal zone usually do not mix with surface waters due to density differences related to temperature.

— Mixing of these layers usually occurs twice each year in temperate lakes and ponds; thus, oxygen enters the profundal zone and nutrients are cycled into the limnetic zone.

Lakes are often classified as oligotrophic or eutrophic, depending on the amount of organic matter produced.

— Oligotrophic lakes are deep, nutrient-poor lakes in which the phytoplankton are not very productive.

  — The water is usually clear and the profundal zone has a high oxygen concentration since little detritus is produced in the limnetic zone to be decomposed.

— Eutrophic lakes are usually shallow, nutrient-rich lakes with very productive phytoplankton.

  — The waters are usually murky due to large phytoplankton populations and the large amounts of detritus being decomposed may result in oxygen depletion in the profundal zone during the summer.

— Mesotrophic lakes have moderate amounts of nutrients and phytoplankton productivity.

Oligotrophic lakes may develop into mesotrophic and then eutrophic lakes over a long period of time.

- Runoff from surrounding terrestrial habitats brings in mineral nutrients and sediments.
- Human activities increase the nutrient content of runoff due to lawn and agricultural fertilizers; municipal wastes dumped into lakes dramatically enriches the nitrogen and phosphorus concentrations, which increases phytoplankton and plant growth.
- Algal blooms and increased plant growth results in more detritus and can lead to oxygen depletion due to increased decomposition.
  - This cultural eutrophication usually makes the water unusable.

Streams and rivers are bodies of water that move continuously in one direction.

- There is a change in structure of these bodies of water from their headwaters (point of origin) to their mouths (where they empty into a larger body of water).
- At the headwaters, the water is cold and clear, carries little sediment, and has few mineral nutrients.
- The channel is narrow with a rocky substrate and the water flows swiftly.
- Near the mouth, water moves slowly and is more turbid due to sediment entering from other streams and erosion; the nutrient content is also higher.
- The channel is usually wider with a silty substrate that has resulted from deposition of silt.

Human activities have greatly affected many streams and rivers.

- Channelization (increases flow rate) and damming (slows flow rate) often change associated ecosystems.
- Pollutants may be taken up by the natural flora and fauna.

Wetlands

A wetland is an area covered by water that supports aquatic vegetation

— Includes a broad range of habitats from periodically flooded regions to soil that is permanently saturated during growing season

— Conditions favor hydrophytes, which are plants specially adapted to grow in water or soil that is periodically anaerobic due to being saturated with water.

  — Cattails, pond lilies, and sedges are examples of hydrophytes.

— Both the hydrology and the vegetation are important determinants in wetland classification.

A wide variety of wetlands has been recognized although they form in one of only three topographic situations:

— Basin wetlands develop in shallow basins ranging from upland depressions to lakes and ponds that have filled in.

— Riverine wetlands develop along shallow and periodically flooded banks of streams and rivers.

— Fringe wetlands are found along coasts of large lakes and seas where rising lake levels or tides cause water to flow back and forth.

  — Fringe wetlands include both freshwater and marine habitats.

  — Marine coastal wetlands are closely linked to estuaries.

Wetlands are among the richest and valuable of biomes.

— A diverse invertebrate community is present which supports a wide variety of birds.

— A variety of herbivorous species consume the algae, detritus, and plants.

— They provide water storage basins that reduce the intensity of flooding.

— They improve water quality by filtering pollutants.

Although many wetlands have been destroyed for agriculture and development, a move to protect the remaining wetlands is underway.

Estuaries

An estuary is the area where a freshwater stream or river merges with the ocean.

— They are often bordered by salt marshes or intertidal mudflats

— Their salinity varies spatially within the estuary from nearly fresh water to ocean water; varies daily in these areas due to rise and fall of tides

— Estuaries are very productive due to nutrients brought in by rivers.

Because of their productivity, estuaries have a diverse flora and fauna.

— Salt marsh grasses, algae, and phytoplankton are the major producers.

— Many species of annelids, oysters, crabs, and fish are also present.

— Many marine invertebrates and fish breed in estuaries or migrate through them to freshwater habitats upstream.

— A large number of water fowl and other semiaquatic vertebrates use estuaries as feeding areas.

Human activities have had a large impact on estuaries.

— Estuaries receive the pollutants dumped into the streams and rivers that feed them.

— Residential and commercial development not only adds to pollution but eliminates some estuaries due to land filling.

— Very little undisturbed estuary habitat remains.

Zonation in marine communities

Similar to freshwater communities, marine communities are distributed according to depth of the water, distance from shore, degree of light penetration, and open water versus bottom.

— A photic zone is present and extends to the depth at which light penetration supports photosynthesis; occupied by phytoplankton, zooplankton, and many fish species.

— The aphotic zone is below the level of effective light penetration and represents a majority of the ocean's volume.

— The intertidal zone is the shallow zone where terrestrial habitat meets the ocean's water.

— The neritic zone extends from the intertidal zone, across the shallow regions, to the edge of the continental shelf.

— The oceanic zone extends over deep water from one continental shelf to another; reaches great depth.

— Pelagic zone refers to open waters of any depth.

— Benthic zone refers to the seafloor.

Intertidal zones

Intertidal zones, where land and sea meet, are alternately submerged and exposed by daily tide cycles.

— Organisms in this zone are exposed to variations in temperature and availability of sea water.

— These organisms are also subjected to the mechanical forces of wave action.

Rocky intertidal zones are vertically stratified and inhabited by organisms that possess structural adaptations that allow them to remain attached in this harsh environment.

- The uppermost zone is submerged only by the highest tides and is occupied by relatively few species of algae, grazing mollusks, and suspension-feeding barnacles.
  - These organisms have various adaptations to prevent dehydration.
- The middle zone is exposed at low tide and submerged at high tide.
  - Many species of algae, sponges, sea anemones, barnacles, mussels, and other invertebrates are found in this area.
  - The diversity is greater here due to the longer time spans this area is submerged.
- Tide pools are often found in the middle zone.
  - These are depressions which are covered during high tide and remain as pools during low tide.
  - Tidepool organisms face dramatic salinity increases as water evaporates at low tide.
- The low intertidal zone is exposed only during the lowest tides and shows the greatest diversity of invertebrates, fishes, and seaweeds.

Sandy intertidal zones and mudflats do not show a clear stratification.

- Wave action continually shifts sand or mud particles; few algae or plants are present.
- Predatory crustaceans and many suspension-feeding worms and clams burrow into the sand or mud and feed when the tide submerges the area.

— Predatory or scavenging crabs and shorebirds often feed on burrowing organisms in these areas.

The diversity of intertidal zones is being reduced by human impact.

— Oil pollution destroys many species.

— Polluted water, old fishing lines, and plastic debris is harmful to most species.

— Recreational use of these areas has greatly reduced the number of beachnesting birds and turtles.

Coral reefs

Coral reefs are found in the neritic zone of warm tropical waters where sunlight penetrates to the ocean floor.

— Sunlight penetration permits photosynthesis and a constant supply of nutrients is provided by currents and waves.

Coral reefs are dominated by the structure of the coral. It is formed by a diverse group of cnidarians that secrete a hard, calcium carbonate external skeleton, which provides a substrate on which other corals, sponges, and algae grow

— Multicellular algae that are encrusted with calcium carbonate add large amounts of limestone to the reefs, as do bryozoans.

The cnidarian coral animals feed on microscopic organisms and organic debris, and they obtain some organic molecules from the photosynthetic products of their symbiotic dinoflagellate algae.

— Coral animals can survive without the dinoflagellates, but their rate of calcium carbon deposition is much slower without them; thus, reef formation by corals depends on this symbiotic association.

Reef communities are very old and grow very slowly. Although many coral reefs are very large, they are delicate and can be severely damaged or destroyed by pollution, human induced damage, or introduced predators (e.g., crown-of-thorns sea star).

Oceanic pelagic biome

The oceanic pelagic biome consists of the open waters far from shore. The area is constantly mixed by circulating ocean currents. Nutrient content is generally lower than in coastal areas because the remains of organisms sink out of the zone to the lower benthic regions.

- — In tropical waters, nutrient content of surface waters is lower than surface waters of temperate oceans because a permanent thermal stratification prevents nutrient exchange with the deep waters.
- — Temperate oceans experience periodic upwellings which carry nutrients from the bottom to the surface.

Plankton is prevalent in this zone.

- — Photosynthetic phytoplankton grow and reproduce in the photic region (top 100m) of this biome.
- — Zooplankton (which graze on the phytoplankton) includes protozoans, copepods, krill, jellies, and larvae of many invertebrates and fishes.
- — Most species stay afloat in this zone through the aid of morphological structures like bubble-trapping spines, lipid droplets, gelatinous capsules, and air bladders.

Nekton (free-swimming animals) are also found in the oceanic pelagic biome. Large squid, fishes, sea turtles, and marine mammals feed in this area. Many fish are adapted to and live in the aphotic region of the pelagic zone.

- — Some have large eyes allowing them to see in dim light, while others have luminescent organs used to attract mates and prey. Many pelagic bird species also feed on fish in this region.

Benthos

Benthos refers to those organisms which inhabit the benthic zone, the ocean bottom below both the neritic and pelagic zones.

— The benthic zone receives nutrients in the form of detritus, which settles into the area from the waters above.

— Light and temperature decline rapidly from shallow, near-shore benthic areas to the ocean's depths.

— The benthic zone substrate may be sand or very fine sediment composed of silt and shells of dead microscopic organisms.

Neritic benthic communities are diverse and productive.

— Many bacteria, fungi, seaweeds, filamentous algae, numerous invertebrates, and fishes are found here.

— Many of the species present are burrowing forms.

— The composition of the community varies with distance from shore, water depth, and bottom composition.

Deep benthic communities living in the abyssal zone are exposed to very different conditions than those under the neritic and pelagic zones.

— Water temperature is continuously cold (3°C), water pressure is extremely high, there is very little (if any) light, and low nutrient concentrations are typical.

— Oxygen is usually present and a fairly diverse community of invertebrates and fishes can be found.

— The deep-sea hydrothermal vent communities are found along midocean ridges in this region.

- These vent communities include chemoautotrophic bacteria as the primary producers in place of photosynthesizing organisms.
- These bacteria obtain energy by oxidizing H2S which forms by a reaction of the hot vent water with dissolved sulfate.
- The bacteria are consumed by a variety of giant polychaete worms,

arthropods, echinoderms, and fishes.

Geographical distribution

The climate and other abiotic factors are important in determining why a particular terrestrial biome is found in an area.

- There are latitudinal patterns of biome distribution over the Earth's surface due to the latitudinal patterns in climate.

Terrestrial biomes are often named for major physical or climatic features and for the predominant vegetation, but each is also characterized by microorganisms, fungi, and animals adapted to that particular environment. Vertical stratification is an important feature of terrestrial biomes and the shape and sizes of plants largely defines the layering.

- In many forests, the layers are the canopy.
- Vertical stratification of a biome's vegetation provides many different habitats for animals

Biomes grade into each other without sharp boundaries and may form ecotones, transitional areas between two communities. Species composition may vary from one location to another within a biome. Biomes are dynamic, and disturbance rather than stability is the rule.

— Disturbance may result in patchy biomes, with several communities represented in one biome.

— Human activities have radically altered the natural patterns of periodic disturbances.

Organismal Ecology

Costs and benefits of homeostasis affect an organism's responses to environmental variation. Organisms survive and reproduce in areas where environmental conditions to which they are adapted are found. The ability to tolerate one factor may be dependent on another factor; for example, many aquatic ectotherms can tolerate reduced oxygen at low temperatures, but not at high temperatures which cause higher metabolic rates.

Organisms faced with a fluctuation in an environmental variable may maintain the homeostasis of their bodies through behavioral and physiological mechanisms (regulators) or by allowing their internal conditions to vary with external conditions (conformers).

— Some species are conformers under certain conditions and become regulators under others

Energy expenditure by the animal is necessary if behavioral or physiological mechanisms are used to maintain homeostasis.

— For organisms to survive and reproduce, the energy "cost" of regulation cannot exceed the benefits of homeostasis.

— Since few organisms are perfect regulators or perfect conformers, a majority of organisms represent a group of evolved strategies which permits them to live in their specific environment.

Allocation principle

The principle of allocation is an important concept for

assessing the responses of organisms to a complex environment.

- This principle holds that each organism has a limited amount of energy that can be allocated for obtaining nutrients, escaping predators, coping with environmental fluctuations, growth, and reproduction.
- Energy expended for one function reduces the amount of energy available for other functions.
- If an organism expends a large amount of energy to maintain homeostasis, less is available for growth, reproduction, and other functions.

The distribution of organisms and the homeostatic mechanisms they possess establish different priorities for energy allocation.

- Conformers living in a stable environment may have more energy available for growth and reproduction; however, their geographic distribution is restricted due to intolerance to environmental change.
- Regulators that allocate a large amount of energy to survive environmental changes have less available for other functions so they grow and reproduce less efficiently; however, they can survive and reproduce over a wider range because they can cope with changing environmental conditions.

An organism's short-term responses to environmental variations operate within a long-term framework. It is important to remember that behavioral, physiological, and morphological responses to environmental change have evolved over evolutionary time to their current levels through natural selection.

- What appear to be short-term adjustments are actually evolutionary adaptations to maintain homeostasis.

— Natural selection also places constraints on the distribution of populations by adapting them to localized environments.

— Organisms adapted to one type of environment may not survive if dispersed a foreign environment or may become extinct if the local environment changes to beyond their tolerance limits.

— The existence of a species in a particular location depends on the species reaching that location and being able to survive and reproduce after getting there.

Since environments vary over space and time, the impact of the variations on a particular species depends on the scale of the variation in relation to the species' overall life history.

Physiological responses: Physiological responses to environmental change are generally slower than behavioral responses although some may occur very rapidly.

— An example of a slow change is when a human moves to an area of less oxygen such as at higher altitudes.

  — After several days to a few weeks the person responds with an increased number of red blood cells being produced.

— An example of a faster change would be when blood vessels in the skin constrict within seconds to reduce loss of body heat when the skin exposed to very cold air.

Physiological adaptation is centered around regulation and homeostasis.

— Both regulators and conformers function most efficiently under certain environmental conditions which are optimal for the organism.

— Efficiency declines both above and below optimal values.

— Physiological responses to changing environments can shift tolerance limits of organisms (acclimation).

- Acclimation is a gradual process and is related to the range of environmental conditions experienced under natural conditions.

Morphological responses: Organisms can react to environmental change with responses that alter body form or internal anatomy. May develop over the lifetime of an individual or across generations. May be forms of acclimation since they are reversible.

- Increase in coat fur or feather density in winter
- Change in coat color between winter and summer

Other morphological changes are irreversible over an individual's lifespan since the environmental variation may have affected growth and differentiation patterns.

- Plants are more morphologically plastic than animals.
- The arrowleaf plant lacks a waxy leaf cuticle when growing in water with submerged leaves (allows absorption of materials directly from the water).
- Arrowleaf plants growing on land have a thick cuticle (reduces water loss) and an extensive root system.

Behavioral responses: Behavioral responses to unfavorable environmental changes can be almost instantaneous in their effects and are easily reversed.

- The quickest response of animals is to move to a new, more favorable location.
- Desert animals return to burrows or move into the shade during the day when heat is most intense.
- Migratory birds migrate to warmer climates to over winter.
- Some animals can modify their immediate environment by cooperative social behavior.

— Honeybees seal the hive during cold periods to conserve heat and cool the hive on hot days by the collective beating of their wings.

— Small mammals may huddle in burrows in cold weather which minimizes the total surface area exposed to cold air, thus reducing heat loss.

Theory of Evolution by Natural Selection

As experienced field naturalists, Darwin and Wallace realized that the Malthusian argument applied with equal force to the whole of the plant and animal kingdoms. The living world is dominated by reproduction, overcrowding, and death. Darwin and Wallace were almost obsessed by this great truth. They appreciated that all organisms possess a potential to multiply that is impossible to realize.

Darwin noted the great fecundity of some species-a single individual of the sea slug Doris may produce 600,000 eggs; the parasitic roundworm Ascaris may produce 64 million-and as an example of the absurd consequences of unimpeded population growth, he used a population of fish, each laying 2,000 eggs: in eight generations this "would cover like a sheet the whole globe, land and water". But he realized that every species "must suffer destruction during some period of its life, and during some season or occasional year, otherwise, on the principle of geometrical increase, its numbers would quickly become so inordinately great that no country could support the product"; and he observed that this was so.

Malthus's essay was concerned with the human population, which, if its intrinsic rate of increase remained unchecked, would, he calculated, be capable of doubling every 25 years and overrunning the planet. Malthus realized that limited resources slowed the growth of populations and placed absolute limits on their size, and that disease, wars, and other disasters also checked population growth.

In one of the examples of population ecology, Darwin counted all the seedlings that emerged from a plot of cultivated ground 3 feet long and 2 feet wide: "Out of 357 no less than 295 were destroyed, chiefly by slugs and insects." Both authors, then, emphasized that most individuals die before they can reproduce and contribute nothing to future generations. Both, though, tended to ignore the important fact that those individuals that do survive in a population may leave different numbers of descendants.

The theory of evolution by natural selection, then, rests on a series of established truths:

— Individuals that form a population of a species are not identical.

— Some of the variation between individuals is heritable-that is, it has a genetic basis and is therefore capable of being passed down to descendants.

— All populations could grow at a rate that would overwhelm the environment; but, in fact, most individuals die before reproduction and most (usually all) reproduce at less than their maximal rate. Hence, each generation, the individuals in a population are only a subset of those that "might" have arrived there from the previous generation.

— Different ancestors leave different numbers of descendants (descendants, not just offspring): they do not all contribute equally to subsequent generations. Hence, those that contribute most have the greatest influence on the heritable characteristics of subsequent generations.

Evolution is the change, over time, in the heritable characteristics of a population or species. Given the above four truths, the heritable features that define a population will inevitably change.

Evolution is inevitable. But which individuals make the disproportionately large contributions to subsequent generations and hence determine the direction that evolution takes? The answer is: those that were best able to survive the risks and hazards of the environments in which they were born and grew; and those who, having survived, were left by their environments most capable of successful reproduction. Thus, interactions between organisms and their environments-the stuff of ecology-lie at the heart of the process of evolution by natural selection.

The philosopher Herbert Spencer described the process as "the survival of the fittest", and the phrase has entered everyday language-which is regrettable. The term fitness is commonly used to describe the success of individuals in the process of natural selection. An individual will survive better, reproduce more, and leave more descendants-it will be fitter-in some environments than in others.

In a given environment, some individuals will survive better, reproduce more, and leave more descendants-they will be fitter-than other individuals. Darwin had been greatly influenced by the achievements of plant and animal breeders-for example, the extraordinary variety of pigeons, dogs, and farm animals that had been deliberately bred by selecting individual parents with exaggerated traits. He and Wallace saw nature doing the same thing-"selecting" those individuals that survived from their excessively multiplying populations-hence the phrase "natural selection". But even this phrase can give the wrong impression. There is a great difference between human and natural selection.

Human selection has an aim for the future-to breed a cereal with a higher yield, a more attractive pet dog, a better hunter, or a cow that will yield more milk. But nature has no aim. Evolution happens because some individuals have survived the death and destruction of the past and reproduced more successfully in the past, not because they were somehow chosen or selected by "Mother Nature" as improvements for the future. Hence, past environments may be said to have selected particular characteristics of individuals that we see in present-day populations. Those characteristics are "suited" to present-day environments only because environments tend to remain the same, or at least change only very slowly.

Darwin and Wallace placed slightly different emphases on the forces that drive evolution. Wallace emphasized the killing forces of physical conditions such as frost, drought, and predators. Darwin laid more emphasis on competition for limited resources and the lethal effects of crowding that result from overpopulation.

Theory of Evolution within Species

The natural world is not composed of a continuum of types of organism each grading into the next; we recognize boundaries between one sort of organism and another. There were features of both plants and animals that were not easily modified by the organisms' immediate environment, and that these "conservative" characteristics were especially useful for classifying organisms.

In flowering plants, the form of the flowers was particularly stable, whereas the size of leaves and stems was much more readily affected by heat and cold, watering and drought, and the giving and withholding of fertilizers. Nevertheless, within what we recognize as species, there is often considerable variation, and some of this is heritable. It is on such intraspecific variation, after all, that plant and animal breeders work. In nature, some of this intraspecific variation is clearly correlated with variations in the environment and represents local specialization.

Darwin called his book On the Origin of Species by Means of Natural Selection, but evolution by natural selection does far more than create new species. Natural selection and evolution occur within species, and we now know that we can study them in action and within our own lifetime. Moreover, we need to study the way that evolution occurs within species if we are to understand the origin of new species.

Geographical Variation

Since the environments experienced by a species in different parts of its range are themselves different (to at least some extent), we might expect natural selection to have favored different variants of the species at different sites. But evolution forces the characteristics of populations to diverge from each other (1) only if there is sufficient heritable variation on which selection can act, and (2) provided that the forces of selection favoring divergence are strong enough to counteract the mixing and hybridization of individuals from different sites. Two populations will not diverge completely if their members (or, in the case of plants, their pollen) are continually migrating between them, mating and mixing their genes.

The sapphire rockcress, Arabis fecunda, is a rare perennial herb restricted to calcareous soil outcrops in western Montana-so rare, in fact, that there are just 19 existing populations separated into two groups ("high elevation" and "low elevation") by a distance of around 100 km. Whether there is local adaptation here is of practical importance: four of the low elevation populations are under threat from spreading urban areas and may require reintroduction from elsewhere if they are to be sustained.

Reintroduction may fail if local adaptation is too marked. Observing plants in their own habitats and checking for differences between them would not tell us if there was local adaptation in the evolutionary sense. Differences may simply be the result of immediate responses to contrasting environments made by plants that are essentially the same. Hence, high and low elevation plants were grown together in a "common garden", eliminating any influence of contrasting immediate environments.

The low elevation sites were more prone to drought: both the air and the soil were warmer and drier; and the low elevation plants in the common garden were indeed significantly more drought tolerant. For example, they had significantly better "water use effficiency" (their rate of water loss through the leaves was low compared to the rate at which carbon dioxide was taken in) as well as being much taller and "broader". (figure 1)

Lengths of shoot taken from this species readily form roots, so that a number of independent rooted plants can be cloned from a single plant taken from the field. Hence, in a careful statistical design, each of four plants taken from each sampling point was represented by five rooted clonal replicates of itself.

The plants spread by sending out shoots along the ground surface (stolons), and the growth of the plants was compared by measuring the lengths of these. In the field, cliff plants formed only short stolons, whereas those of the pasture plants were long. In the experimental garden, these differences were maintained, even though the sampling points were typically only around 30 m apart-certainly within the range of pollen dispersal between plants. Indeed, the gradually changing environment along the transect was matched by a gradually changing stolon length, presumably with a genetic basis, since it was apparent in the common garden. Even here, the forces of selection seem to outweigh the mixing forces of hybridization.

On the other hand, it would be quite wrong to imagine that local selection always overrides hybridization-that all species exhibit geographically distinct variants with a genetic basis. For example, in a study of Chamaecrista fasciculata, an annual legume from disturbed habitats in eastern North America, plants were grown in a common garden that were derived from the "home" site or were transplanted from distances of 0.1, 1, 10, 100, 1,000, and 2,000 km.

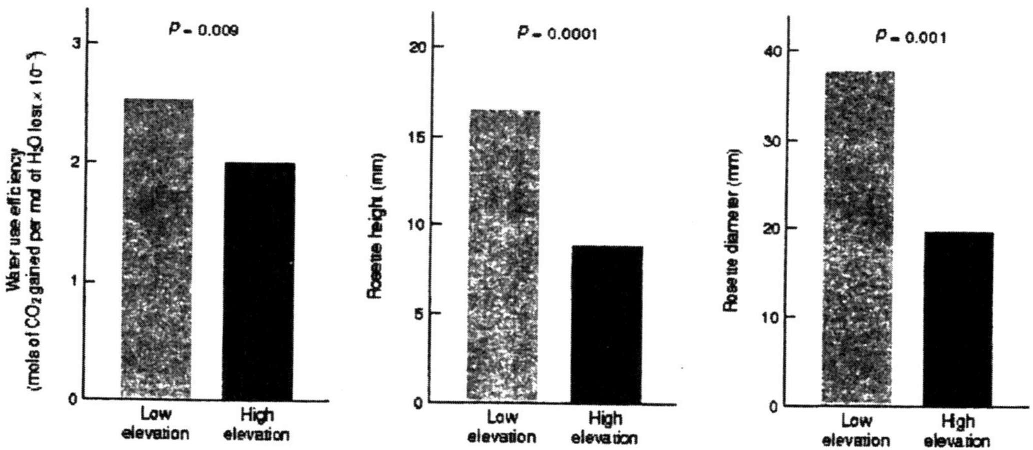

Figure 1. When plants of the rare sapphire rockcress from low elevation (drought prone) and high elevation sites were grown together in a common garden, there was local adaptation; those from the low elevation site had significantly better water use efficiency as well as having both taller and broader rosettes.

The study was replicated three times-in Kansas, Maryland, and northern Illinois. Five characteristics were measured: germination, survival, vegetative biomass, fruit production, and the number of fruit produced per seed planted; but for all characters in all replicates there was little or no evidence for local adaptation except at the very furthest spatial scales (Figure 2). There is "local adaptation"- but it's clearly not that local.

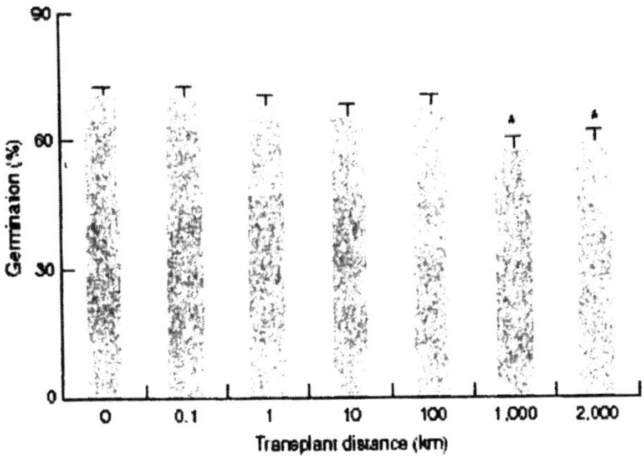

Figure 2.

Figure 2 shows, percentage germination of local and transplanted Chamaecrista fasciculata populations to test for local adaptation along a transect in Kansas. Data for 1995 and 1996 have been combined because they do not differ significantly. Populations that differ from the home population at $P < 0.05$ are indicated by an asterisk. Local adaptation occurs at only the largest spatial scales.

We can also test whether organisms have evolved to become specialized to life in their local environment in reciprocal transplant experiments, comparing their performance when they are grown "at home" (i.e., in their original habitat) with their performance "away" (i.e., in the habitat of others). It can be difficult to detect the local specialization of animals by transplanting them into each other's habitat: if they do not like it, most species will run away.

But invertebrates like corals and sea anemones are sedentary, and some can be lifted from one place and established in another. The sea anemone Actinia tenebrosa is found in pools on headlands around the coast of New South Wales, Australia. Ecologists chose three colonies on headlands within 4 km of each other on which the anemone was abundant. Within each colony, they selected three transplant sites (each 3-5 m long) and at each set aside three 1 m wide strips-two to receive anemones from the away sites and one to receive "transplanted" individuals from the home site itself.

They cleared the experimental sites of all the anemones present and transplanted anemones into them. The anemone multiplies clonally by producing broods of asexual juveniles. The number of juveniles brooded per adult was used as a measure of the performance of the anemones in the various pools (home and away). The proportion of adults that were found brooding 11 months later is shown in Table 1. Anemones originally sampled from Green Island were rather successful in brooding young after being transplanted both home and away and did not show any specialization to their home environment.

Table 1. A reciprocal transplant experiment of the sea anemone Actinia tenebrosa.

	Transplanted to sites at:			
Site of origin		Green Island	Salmon Point	Strickland Bay
Green Island	a	*0.42*	0.68	0.78
	b	*0.80*	0.63	0.75

	c	*0.67*	0.62	0.61
Salmon Point	a	0.11	*0.42*	0.13
	b	0.18	*0.43*	0.28
	c	0.00	*0.50*	0.40
Strickland Bay	a	0.11	0.06	*0.33*
	b	0.00	0.06	*0.27*
	c	0.04	0.20	*0.27*

a, b, and c are the three replicate sites in each colony. In each case the proportion of adults that were found brooding young is shown. Transplants back to the home sites are shown in italics print.

However, in all the other transplant experiments a greater proportion of anemones brooded young at home than at away sites: strong evidence of evolved local specialization. In later experiments ecologists lifted anemones from a variety of sites as before, but they then kept them for a period to acclimate at a common site before transplanting them in a reciprocal experiment. This more severe test convincingly confirmed the results in Table 1. Another reciprocal transplant experiment was carried out with white clover (Trifolium repens), which forms clones in grazed pastures. Individual clones differ in features such as the pattern of white markings on the leaves, ability to release hydrogen cyanide when damaged or bitten, and susceptibility to various diseases.

To determine whether the characteristics of individual clones matched local features of their environment, Turkington and Harper removed plants from marked positions in the field and multiplied them into clones in the common environment of a greenhouse. They then transplanted samples from each clone into the place in the sward of vegetation from which it had originally been taken, and also to places from where all the others had been taken. The plants were allowed to grow for a year before they were removed, dried, and weighed.

The mean weight of clover plants transplanted back into their home sites was 0.89 g, but at away sites it was only 0.52 g, a statistically highly significant difference. The clover plants studied were not random samples but had been chosen from patches dominated by four different species of grass. Hence, in a second experiment, clonal samples of the different clovers were planted into dense experimental plots of the four grasses. Again the clovers were removed after 12 months, dried, and weighed. The mean yield of clovers grown with their original neighbor grass was 59.4 g; the mean yield with "alien" grasses was 31.9 g, again a highly significant difference.

Thus, both clover experiments provide strong direct evidence that clover clones in the pasture had evolved to become specialized, such that they tended to perform best (make most growth) in their local environment and with their local neighbors. In most of the examples so far, geographic variants of species have been identified, but the selective forces favoring them have not. This is not true of the next example. The guppy (Poecilia reticulata), a small freshwater fish from northeastern South America, has been the material for a classic series of evolutionary experiments.

In Trinidad, many rivers flow down the northern range of mountains and are subdivided by waterfalls that isolate fish populations above the falls from those below. Guppies are present in almost all these water bodies, and in the lower waters they meet various species of predatory fish that are absent higher up the rivers. The populations of guppies in Trinidad differ from each other in almost every feature that biologists have examined. Forty-seven of these traits tend to vary in step with each other (they covary) and with the intensity of the risk from predators. This correlation suggests that the guppy populations have been subject to natural selection from the predators. But the fact that two phenomena are correlated does not prove that one causes the other. Only controlled experiments can establish cause and effect.

Females are dull and dowdy and inconspicuous. Whenever we study natural selection in action, it becomes clear that compromises are involved. For every selective force that favors change, there is a counteracting force that resists the change. Color in male guppies is a good example. Female guppies prefer to mate with the most gaudily decorated males—but these are more readily captured by predators because they are easier to see. This sets the stage for some revealing experiments on the ecology of evolution. Guppy populations were established in ponds in a greenhouse and exposed to different intensities of predation.

In an experiment, 200 guppies were moved from a site far down the Aripo River where predators were common and introduced to a site high up the river where there were neither guppies nor predators. The transplanted guppies thrived in their new site, and within just 2 years the males had more and bigger spots of more varied color. The females' choice of the more flamboyant males had had dramatic effects on the gaudiness of their descendants, but this was only because predators did not reverse the direction of selection. The speed of evolutionary change in this experiment in nature was as fast as that in artificial selection experiments in the laboratory.

There was plenty of overpopulation and there was considerable genetic variation in the populations upon which natural selection could act. The guppies transplanted into the nearly predatorfree environment evolved in other respects too. The females were larger and older at maturity, they produced fewer but bigger offspring, and they began to lose the habit of moving together in schools-behavior that defends them against predators.

Natural selection changes the character of a population by sifting out and eliminating much of its variation and leaving behind a residue for future generations with a narrower range and more restricted potential. This is commonly pictured as a force that drives populations toward a peak of adaptation-of a perfect match between organism and environment. This is an optimist's view (Figure 3a). An alternative picture of natural selection is that it forces populations into an ever-narrowing rut of overspecialization an ever-deepening trap (Figure 3b). This pessimist's view emphasizes how the effects of natural selection are to limit and constrain, and that the specialization of species means that they risk extinction when the environment changes.

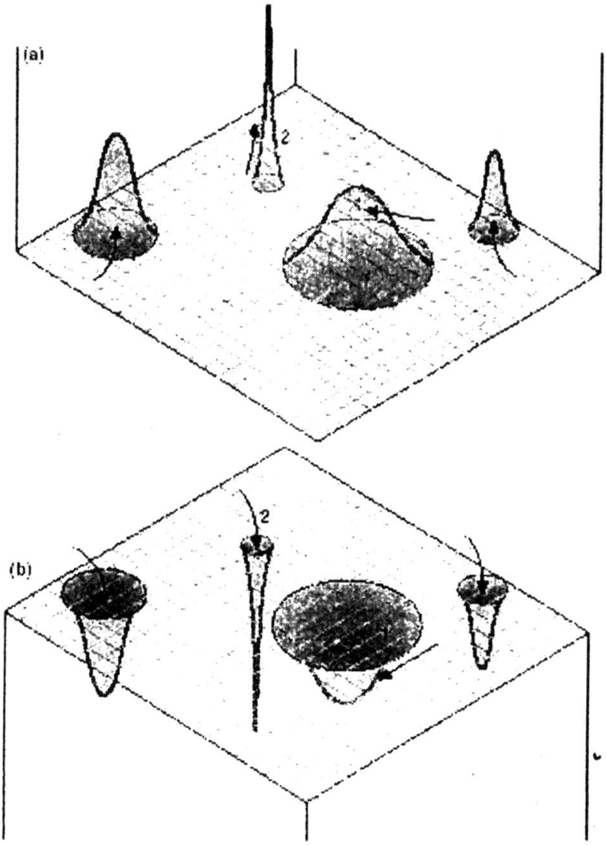

Figure 3. Two models of natural selection in action: (a) an optimistic and (b) a pessimistic interpretation of natural selection.

It is easy to see that a population of plants faced with repeated drought is likely to evolve a tolerance of water shortage, and an animal repeatedly faced with cold winters is likely to evolve habits of hibernation or a thick protective coat. But droughts do not become any less severe as a result, nor winters milder.

figure 3 shows, In both (a) and (b) a horizontal plane is drawn to represent two dimensions of a range of environmental conditions. On both diagrams ellipses are drawn to show the range of this variation that is tolerated by four populations of organisms. (a) In the optimistic view, the vertical scale is a measure of the range of fitness of the organisms in the populations and natural selection is shown as arrows driving the population to ever higher "adaptive" peaks. Population 1 is highly variable and tolerates a wide range of conditions.

Natural selection is relatively weak. Population 2 is a very uniform population, and natural selection is fierce and driving the population to a very high degree of specialization and local fitness. (b) In the pessimistic view, natural selection is shown as arrows driving the populations into ruts, troughs, and pits. The vertical downward axis is a measure of the intensity of specialization. The highly variable population with weak selection is rather safe if the environment changes but population 2 is at extreme risk of extinction. The two depictions should not be seen as one right and one wrong impression of natural selection in action, but rather as two views of the same truths.

Physical conditions are not heritable: they leave no descendants, and they are not subject to natural selection. But the situation is quite different when two species interact: predator on prey, parasite on host, competitive neighbor on neighbor. Natural selection may select from a population of parasites those forms that are more efficient at infecting their host. But this immediately sets in play forces of natural selection that favor more resistant hosts. As they evolve they put further pressure on the ability of the parasite to infect.

Host and parasite are then caught in never-ending reciprocating selection. A result is that both host and parasite become increasingly specialized-caught in an ever-deepening rut. Eventually only a specialized form of the parasite can infect and can do so only on a highly specialized form of the host.

Speciation Ecology

Natural selection can force populations of plants and animals to change their character-to evolve. But none of the examples we have considered has involved the evolution of a new species. Indeed Darwin's On the Origin of Species is about natural selection and evolution but is not really about the origin of species! Those who studied the evolution of melanism in the peppered moth named the black and normal forms forma carbonaria and forma typica: they classified them as forms within a species, not as different species.

A species is what a competent taxonomist regards as a species! Darwin himself regarded species (like genera) as "merely artificial combinations made for convenience". On the other hand, back in the 1930s, two American biologists, Ernst Mayr and Theodosius Dobzhansky, proposed an empirical test that could be used to decide whether two populations were part of the same or of two different species. They recognized organisms as being members of a single species if they could, at least potentially, breed together in nature to produce fertile offspring. They called a species tested and defined in this way a biospecies.

If the members of two populations are able to hybridize and their genes are combined and reassorted in their progeny, then natural selection can never make them truly distinct. Although natural selection may tend to force a population to evolve into two or more distinct forms, sexual reproduction and hybridization mix them up again. Two parts of a population can evolve into distinct species only if some sort of barrier prevents gene flow between them; they might, for example, be isolated on different islands. While isolated from each other they may then evolve and become so different that, if they meet again, they can no longer hybridize and their populations can no longer exchange genes. They are now different biospecies.

Figure 4 illustrates this process. A uniform species with a large range (1) differentiates (2) into local forms, varieties, or subspecies, which (3) become genetically isolated from each other, for example, separated by geographical barriers or dispersed onto different islands. After evolution in isolation they may meetagain (4) when they are unable to hybridizeand have become true biospecies. Differences that are particularly effective in keeping newly evolved species distinct are different rituals of courtship, different signals of attraction between the sexes, and in flowering plants, different species of insect pollinator. It may sometimes happen that hybrids form between two evolving species but their parental chromosomes have become so different that they fail to pair at meiosis: the hybrids are then sterile (for example, the horse-donkey hybrid is the sterile mule).

The evolution of species and the balance between natural selection and hybridization are illustrated by the extraordinary case of two species of sea gull. The lesser black-backed gull (Larus fuscus) originated in Siberia and colonized progressively to the west, forming a chain or cline of different forms, spreading from Siberia to Britain and Iceland. The neighboring forms along the cline are distinctive, but they hybridize readily in nature. Neighboring populations are therefore regarded as part of the same species and taxonomists give them only "subspecific" status (e.g., Larus fuscus graellsii, Larus fuscus fuscus).

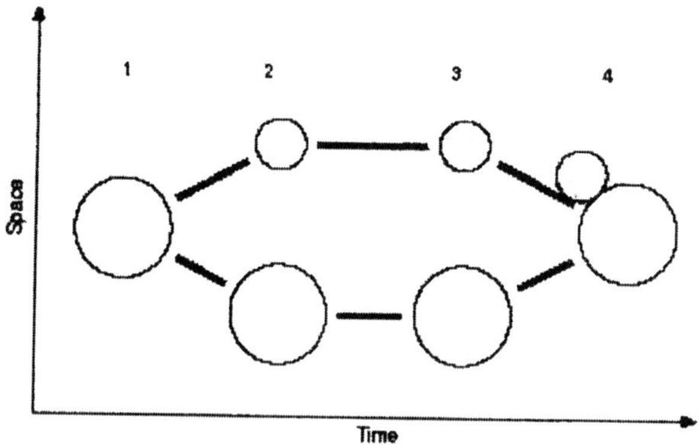

Figure 4. The role of isolation in the evolution of species.

Populations of the gull have, however, also spread east from Siberia, again forming a cline of freely hybridizing forms. Together, the populations spreading east and west encircle the northern hemisphere. They meet and overlap in northern Europe. There, the eastward and westward clines have diverged so far that it is easy to tell them apart, and they are recognized as two different species, the lesser black-backed gull and the herring gull. Moreover, the two species do not hybridize: they have become true biospecies.

In this remarkable example we can see how two distinct species have evolved from one primal stock, and that the stages of their divergence remain frozen in the cline that connects them. But it is where a population becomes split into completely isolated populations, dispersed onto different islands for example, that they most readily diverge into distinct species.

Islands and Speciation

The most celebrated example of evolution and speciation on islands is the case of Darwin's finches in the Galapagos archipelago. The Galapagos are volcanic islands isolated in the Pacific Ocean about 1000 km west of Equador and 750 km from the island of Cocos, which is itself 500 km from

Central America. At more than 500 m above sea level the vegetation is open grassland. Below this is a humid zone of forest that grades into a coastal strip of desert vegetation with some endemic species of prickly pear cactus (Opuntia).

Fourteen species of finch are found on the islands, and there is every reason to suppose that these evolved from a single ancestral species that invaded the islands from the mainland of Central America. In their remote island isolation, the Galapagos finches have radiated into a variety of species in groups with contrasting ecologies. Members of one group, including Geospiza fuliginosa and G. fortis, have strong bills and hop and scratch for seeds on the ground. Geospiza scandens has a narrower and slightly longer bill and feeds on the flowers and pulp of the prickly pears as well as on seeds.

Finches of a third group have parrot-like bills and feed on leaves, buds, flowers, and fruits, and a fourth group with a parrot-like bill (Camarhynchus psittacula) has become insectivorous, feeding on beetles and other insects in the canopy of trees. A so-called woodpecker finch, Camarhynchus (Cactospiza) pallida, extracts insects from crevices by holding a spine or a twig in its bill. Yet a further group includes a species (Certhidea olivacea) that, rather like a warbler, fits around actively and collects small insects in the forest canopy and in the air.

Populations of ancestor species became reproductively isolated, most likely after chance colonization of different islands within the archipelago, and evolved separately for a time. Subsequent movements between islands may have brought nonhybridizing biospecies together, and subsequently these have evolved to fill different niches. When individuals from different species compete, natural selection may act to favor those individuals that compete least with members of the other species.

An expected consequence is that among a group of closely related species, such as Darwin's finches, differences in feeding and other aspects of their ecology are likely to become enhanced with time. The entire process of evolutionary divergence of these species appears to have happened in less than 3 million years. Very rarely, hybridization occurs among species that have similar ecologies. However, these occasions are so rare that the species are true or emerging biospecies. We need to remember that the origin of a species is normally a process rather than an event.

For the formation of a new species, like the boiling of an egg, there is some freedom to argue about when it is completed! The evolutionary relationships among the various Galapagos finches have been traced by molecular techniques (analyzing variation in "microsatellite" DNA). These accurate modern tests confirm the long-held view that the family tree of the Galapagos finches radiated from a single trunk (i.e., was monophyletic) and also provides strong evidence that the warbler finch was the first to split off from the founding group and is likely to be the most similar to the original colonist ancestors.

The flora and fauna of many other archipelagos show similar examples of great richness of species with many local endemics (i.e., species known only from one island or area). The Hawaiian Islands are home to an extraordinary diversity of picture-winged fruit flies and closely related species of honey creeper that have diverged in feeding habit and bill shape remarkably like the Galapagos finches.

Lizards of the genus Anolis have evolved a kaleidoscopic diversity of species on the islands of the Caribbean; and isolated groups of islands, such as the Canaries off the coast of North Africa, are treasure troves of endemic plants. These unusual and often rich communities may pose particular problems for the applied ecologist. Invaders onto marine islands may become isolated from other parts of their population and are then free to diverge under natural selection and become different species. But there are other kinds of "islands" in which colonists can also become genetically isolated from the rest of a population.

Mountains isolate valleys from each other and valleys isolate mountains. A few individuals that chance to be dispersed to a habitable site in a mountain range can form the nucleus of an expanding new species. Its character will have been colored by the particular genes that were represented among the colonists-unlikely to be a perfect sample of the parent population. What natural selection can do with this founder population is limited by what is in its limited sample of genes (plus occasional rare mutations).

Indeed much of the deviation among populations isolated on islands appears to be due to a founder effect-the chance composition of the pool of founder genes puts limits and constraints on what variation there is for natural selection to act upon. The evolutionary biologist's understanding of island patterns depends on a thorough appreciation of ecological processes such as dispersal and interspecific competition.

Effects of Climatic Change on Species Distribution

Changes in climate, particularly during the ice ages of the Pleistocene, bear a lot of the responsibility for the present patterns of distribution of plants and animals. As climates have changed, species populations have advanced and retreated, have been fragmented into isolated patches, and may have then rejoined. Much of what we see in the present distribution of species represents phases in a recovery from past climatic change.

Modern techniques for analyzing and dating biological remains (particularly buried pollen) are beginning to allow us to detect just how much of the present distribution of organisms is a precise, local-evolved match to present environments, and how much is a fingerprint left by the hand of history. For most of the past 2-3 million years the

Earth has been very cold. Evidence from the distribution of oxygen isotopes in cores taken from the deep ocean floor shows that there may have been as many as 16 glacial cycles in the Pleistocene, each lasting for up to 125,000 years.

Each cold (glacial) phase may have lasted for as long as 50,000-100,000 years, with brief intervals of only 10,000-20,000 years when the temperatures rose to, or above, those of today. In this case, present floras and faunas are unusual, having developed at the warm end of one of a series of unusual catastrophic warm periods. During the 20,000 years since the peak of the last glaciation, global temperatures have risen by about 8°C. The analysis of buried pollen-particularly of woody species, which produce most of the pollen-can show how vegetation has changed during this period. As the ice retreated, different forest species advanced in different ways and at different speeds.

For some, like the spruce of eastern North America, there was displacement to new latitudes; for others, like the oaks, the picture was more one of expansion. We do not have such good records for the postglacial spread of the animals associated with the changing forests, but it is at least certain that many species could not have spread faster than the trees on which they feed. Some of the animals may still be catching up with their plants, and tree species are still returning to areas they occupied before the last ice age! It is quite wrong to imagine that our present vegetation is in some sort of equilibrium with the present climate. Even in regions that were never glaciated, pollen deposits record complex changes in distributions: range shifts to new regions, large changes in population size, both up and down, and in the mountains of the Sheep Range, Nevada, shifts in elevation range varying in amount and even direction.

The species composition of vegetation has continually been changing and is almost certainly still doing so. The records of climatic change in the tropics are far less complete than those for temperate regions. Many believe, though, that during cooler, drier glacial periods, the tropical forests retreated to smaller patches, surrounded by a sea of savanna. Support for this comes from the present-day distribution of species in the tropical forests of South America. There, particular "hot-spots" of species diversity are apparent, and these are thought to be likely sites of forest refuges during the glacial periods, and sites too, therefore, of increased rates of speciation.

On this interpretation, the present distributions of species may again be seen as largely accidents of history (where the refuges were) rather than precise matches between species and their differing environments. Evidence of changes in vegetation that followed the last retreat of the ice hint at the likely consequences of the global warming (maybe 3°C in the next 100 years) that is predicted to result from continuing increases in "greenhouse" gases in the atmosphere. But the scales are quite different. Postglacial warming of about 8°C occurred over around 20,000 years, and changes in the vegetation failed to keep pace even with this.

But current projections for the 21st century require range shifts for trees at rates of 300-500 km per century compared to typical rates in the past of 20-40 km per century. It is striking that the only precisely dated extinction of a tree species in the Quaternary, that of Picea critchfeldii, occurred around 15,000 years ago at a time of especially rapid postglacial warming. Clearly, future even more rapid change could result in extinctions of many additional species.

Continental Drift Effects on the Ecology of Evolution

The patterns of species formation that occur on islands appear on an even larger scale in the evolution of genera and families across continents. Many curious distributions of organisms between continents seem inexplicable as the result of dispersal over vast distances. Biologists, especially Wegener, met outraged scorn from geologists and geographers when they argued that it must have been the continents that had moved rather than the organisms that had dispersed. Eventually, however, measurements of the directions of the Earth's magnetic fields required the same, apparently wildly improbable, explanation and the critics capitulated.

The discovery that the tectonic plates of the Earth's crust move and carry the migrating continents with them reconciles geologist and biologist. While major evolutionary developments were occurring in the plant and animal kingdoms, their populations were being split and separated, and land areas were moving across climatic zones. This was happening while changes in temperature were occurring on a vastly greater scale than the glacial cycles of the Pleistocene episode. The established drift of the continents answers many questions in the ecology of evolution. The curious world distribution of large flightless birds is one example (Figure 5a).

The presence of the ostrich in Africa, the emu in Australia, and the very similar rhea in South America could scarcely be explained by dispersal of some common flightless ancestor. Now, techniques of molecular biology make it possible to analyze the time at which the various flightless birds started their evolutionary divergence (Figure 5b). The tinamous seem to have been the first to diverge and became evolutionarily separate from the rest, the ratites. Australasia next became separated from the other southern continents, and from the latter, the ancestral stocks of ostriches and rheas were subsequently separated when the Atlantic opened up between Africa and South America.

Back in Australasia, the Tasman Sea opened up about 80 million years ago and ancestors of the kiwi are thought to have made their way, by island hopping, about 40 million years ago across to New Zealand, where divergence into the present species happened relatively recently. The unraveling of this particular example implies the early evolution of the property of flightlessness and only subsequently the isolation of the different types between the emerging continents.

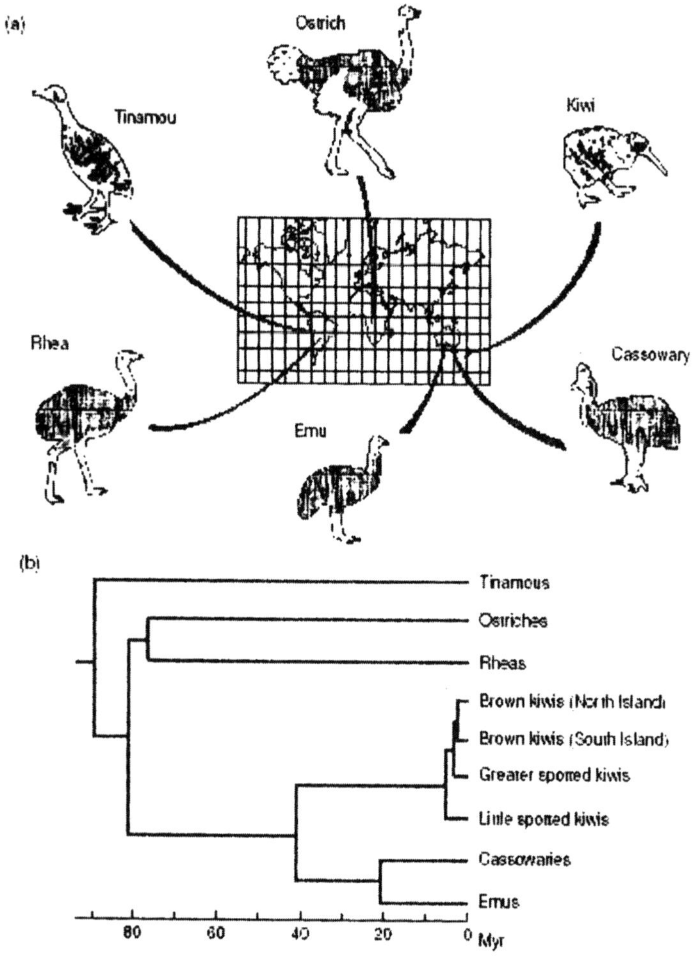

Figure 5. (a) The distribution of terrestrial flightless birds. (b) The phylogenetic tree of the flightless birds and the estimated times (million years ago) of their divergence.

Interpreting the Evolution Results

Flightlessness did not evolve independently on the different continents. However, there are many examples of organisms that have evolved in isolation from each other and then converged on remarkably similar forms or behavior. Such similarity is particularly striking when similar roles are played by structures that have quite different evolutionary origins-that is, when the structures are analogous but not homologous. When this occurs, it is termed convergent evolution.

Bird and bat wings are a classic example (Figure 6). the wings of birds are analogous (not homologous). They are structurally different—the bird wing is supported by digit number 2 and is covered with feathers; the bat wing is supported by digits 2–5 and is covered with skin. Such convergence is direct evidence of the power of evolutionary forces to shape the same form from quite different starting material. The French geneticist Jacob said that evolution was like "tinkering". It did not create ideal forms from ideal beginnings rather it tinkered together what it could from what was available at the time. Further examples show parallels in the evolutionary pathways of ancestrally related groups occurring after they were isolated from each other.

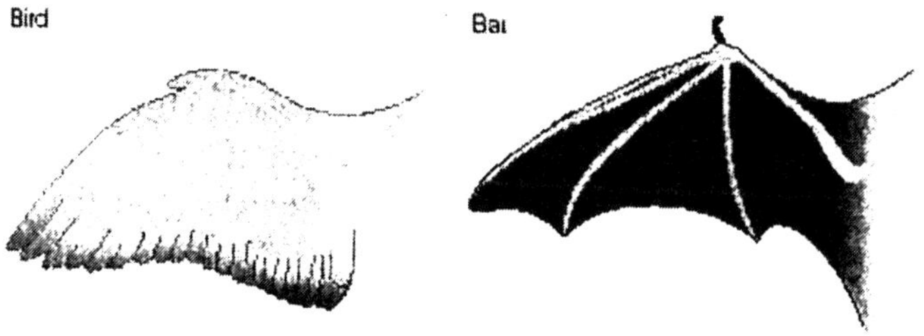

Figure 6. Convergent evolution

The classic example is provided by the placental and marsupial mammals. Marsupials arrived on what would become the Australian continent in the Cretaceous period, when the only other mammals present were the curious egg-laying monotremes (now represented only by the spiny anteaters and the duckbilled platypus). An evolutionary process of radiation then occurred among the Australian marsupials that in many ways accurately paralleled what was occurring among the placental mammals on other continents. It is hard to escape the view that the environments of placentals and marsupials contained ecological pigeonholes (niches) into which the evolutionary process has neatly "fitted" ecological equivalents.

In contrast to convergent evolution, however, the marsupials and placentals started to diversify from a common ancestral line, and both inherited a common set of potentials and constraints. When we marvel at the diversity of complex specializations by which organisms match their varied environments there is a temptation to regard each case as an example of evolved perfection. But there is nothing in the process of evolution by natural selection that implies perfection. For example, no population of organisms can contain all the possible genetic variants that might exist and might influence fitness.

The evolutionary process works on the genetic variation that is available. It favors only those forms that are fittest from among the range of variety available, and this may be a very restricted choice. The very essence of natural selection is that organisms come to match their environments by being "the fittest available" or "the fittest yet"- they are not "the best imaginable". It is particularly important to realize that past events on the Earth can have profound repercussions on the present. Our world has not been constructed by taking each organism in turn, testing it against each environment, and molding it so that every organism finds its perfect place. It is a world in which organisms live where they do for reasons that are often, at least in part, accidents of history.

2

Environmental Pollution

The atmosphere can be a significant source of pollution to surface water is a relatively recent idea, first demonstrated for acid rain, and later for mercury, PCBs, and nutrients such as nitrogen (N) and phosphorus (P). Unfortunately, it is often assumed that rain water is free of pollutants. For instance, it is sometimes assumed that pollutants in urban runoff are picked up by clean precipitation running off dirty surfaces. Yet it is possible that the rain already contains some of the pollutants, such as phosphorus, nitrogen, mercury, pesticides, and PCBs.

In the case of urban runoff, the creation of impervious surfaces alone may create a nonpoint source (NPS) pollution problem for surface water, even without considering the watershed activities that contribute pollutants, such as lawn care, pet feces, eroded soil, and vegetative litter. The importance of atmospheric loading will vary greatly depending on the pollutant and the nature of the watershed. In urbanized and agricultural watersheds, nutrient loading from the atmosphere may be negligible. But in the same watersheds, the atmosphere may be the main source of toxic pollutants, such as PCBs and mercury.

There are two situations where atmospheric deposition may be especially important sources of NPS pollution to surface water. First, lakes with a small watershed to lake surface area ratio can receive a large proportion of their loading from the atmosphere. For example, a study of Lake Mille Lacs suggests that precipitation (wet and dry fall) may contribute approximately 48 percent of the annual phosphorus loading to the lake. (Lake Mille Lacs occupies 53 percent of its total watershed area.) Similarly, airborne dust is thought to deliver the majority of phosphorus loading to Lake Superior. Second, some pollutants may be primarily delivered by the atmosphere even when there is significant human activity in the watershed.

For instance, the geological source material in most watersheds does not contain a significant source of mercury. When mercury is found in a waterbody it is most likely a result of atmospheric deposition. In addition, environmentally significant levels often accumulate in soils due to atmospheric deposition. If soil is eroded or inundated (say, through impoundment), there may be significant increases in mercury contamination to aquatic systems in the watershed. The development of impervious surfaces (paving, etc.) and storm sewers has the effect of increasing the efficacy of transport to surface water of deposition to the watershed, not only for mercury but also for other airborne pollutants.

Atmospheric Deposition

The emission source to the atmosphere may well be a point source such as an emission stack. It is worth pointing out that even if modeling or measurement studies verify a direct relationship between a point source of air emissions and deposition to a water body, water managers may still consider that source of pollution to be nonpoint, because it is delivered by the atmosphere.

Air managers identify three basic categories of emission: point sources, area sources, and mobile sources.

Each category is further subdivided into subcategories. Point sources are permanently fixed stacks of known diameter, elevation, temperature, and exit velocity. Area sources include windblown dust from stockpiles or tilled fields, fugitive emissions from a landfill or the numerous valves and connections at a refinery, and forest fires. Mobile sources are divided into on-road sources such as traffic emissions and dust from unpaved roads, and off-road sources such as lawn mowers, portable generators, chain saws, and snowmobiles.

Wet Deposition: Pollutants in the atmosphere can be scavenged by precipitation or act as condensation nuclei for precipitation formation and thereby be deposited to surface water and land in the form of rain or snow.

Dry Deposition: Particles in the air are deposited onto surface water and land surfaces at a rate that depends on the particle size, wind speed, and other factors. Gaseous pollutants can also be deposited to water and land.

Indirect and Direct Deposition: Air pollutants are not only deposited directly to the surface of waterbodies, but are also deposited to watersheds and then enter surface waters indirectly, through storm water runoff, tributaries, and groundwater seepage. Where the watershed is large relative to the open water, indirect loading can exceed direct loading.

Volatization

Previously deposited gaseous and semivolatile chemicals, such as mercury and PCBs, can be re-emitted to the atmosphere as the result of many factors, including chemical reactions and changes in temperature or wind speed.

Effects of Air Pollution

Any change in the physics or chemistry of the atmosphere

can negatively affect surface water. For example, depletion of stratospheric ozone could increase the damage to aquatic life from increased UV radiation. Global warming is projected to virtually eliminate the cold water fishery in Minnesota, while simultaneously reducing the duration of ice-cover and therefore winterkills. A wide variety of materials are deposited from the atmosphere that can affect the surface water.

Some airborne materials are toxic (e.g. mercury, PCBs, lead, dioxin), some are nutrients (e.g., phosphorus and nitrogen), and some interact with other pollutants (e.g., calcium carbonate in windblown soil can neutralize acid rain, or sulfate deposition may stimulate the methylation of mercury in low-sulfate systems). The following is a description of the different types of changes in the atmosphere that can affect surface water.

Efects of Greenhouse Gases

Scientists believe that emissions of certain gases to the atmosphere are causing warming and possibly other changes in the climate. The greenhouse gases include the naturally occurring compounds carbon dioxide (CO_2), methane (CH_4), and nitrous oxide (N_2O). Humans also release synthetic greenhouse gases that contribute significantly to climate change (chlorofluorocarbons (CFCs), hydrofluorocarbons (HFCs), and perfluorocarbons (PFCs). Carbon dioxide is released to the atmosphere when solid waste, fossil fuels (oil, natural gas, and coal), and wood and wood products are burned.

Methane is emitted during the production and transport of coal, natural gas, and oil. Methane emissions also result from the decomposition of organic wastes in municipal solid waste landfills, and the raising of livestock. Nitrous oxide is emitted during agricultural and industrial activities, as well as during combustion of solid waste and fossil fuels. Greenhouse gases that are not naturally occurring include byproducts of foam production, refrigeration, and air conditioning called chlorofluorocarbons (CFCs), as well as hydrofluorocarbons (HFCs) and perfluorocarbons (PFCs) generated by industrial processes.

Each greenhouse gas differs in its ability to absorb heat in the atmosphere. HFCs and PFCs are the most heat-absorbent. Methane traps over 21 times more heat than carbon dioxide, and nitrous oxide absorbs 270 times more heat than carbon dioxide. Global warming has already caused significant reductions in the duration of ice cover in Minnesota. Models show that winterkills will get rarer. As summer temperatures rise, summer kills will become more common and in some lakes cold water fisheries will shift to warm water. There will be harder to predict effects on other temperature- and CO_2-sensitive processes, such as mercury methylation and plant growth. Since chemical reaction rates, and the growth rates of bacteria, plants, and cold-blooded animals are all highly dependent on temperature, there may be many unanticipated effects of global warming.

When chlorofluorocarbons (CFCs) reach the stratosphere, the ultraviolet radiation from the sun causes them to break apart and release chlorine atoms which react with ozone, starting chemical cycles of ozone destruction that deplete the ozone layer. One chlorine atom can break apart more than 100,000 ozone molecules. Other chemicals that damage the ozone layer include methyl bromide (used as a pesticide), halons (used in fire extinguishers), and methyl chloroform (used as a solvent in industrial processes).

As methyl bromide and halons are broken apart, they release bromine atoms, which are 40 times more destructive to ozone molecules than chlorine atoms. Reductions in stratospheric ozone levels lead to higher levels of UVB reaching the Earth's surface. Studies have shown that in the Antarctic, the amount of UVB measured at the surface can double during the annual ozone hole. Another study confirmed the relationship between reduced ozone and increased UVB levels in Canada during the past several years.

Ozone levels vary by season and latitude. In the middle latitudes (most of the populated world), ozone levels have fallen about 10% during the winter and 5% in the summer. Since 1979, they have fallen about 5% per decade when averaged over the entire year. Depletion is generally worse at higher latitudes, i.e. further from the Equator. In the marine environment, solar UVB radiation has been found to cause damage to early developmental stages of fish, shrimp, crab, amphibians and other animals. The most severe effects are decreased reproductive capacity and impaired larval development. Even at current levels, solar UVB radiation is a limiting factor in some systems. It is uncertain what effect enhanced UVB radiation would have on the Minnesota environment.

Mercury vapor emissions from combustion sources result in ambient air concentrations below those of concern for direct human health effects through inhalation. Once in the atmosphere, mercury vapor is slowly converted to an ionized form that is water soluble, and subject to wash out in precipitation. Its concentration in rain is usually above the ambient surface water quality standard of 6.9 nanograms per liter (ng/L) (1.3 ng/L in the Lake Superior basin). Some proportion (usually between 1 to 20%) of this mercury is converted to methyl mercury by sulfate-reducing bacteria in the aquatic system or its watershed.

Methyl mercury is bioaccumulated to a great degree in the aquatic food chain. Methylation rates appear to be higher in wetlands than other environments by one or two orders of magnitude. Mercury is probably the most pervasive type of atmospheric NPS pollution in Minnesota, causing fish consumption restrictions on over 90 percent of the lakes tested in the state.

Acid Rain

Sulfuric Acid: Sulfuric acid presents the potential for acidification of surface water, although there is no known permanent damage in Minnesota. There is evidence that increased loading of sulfate stimulates the growth of bacteria that convert sulfate to sulfide in wetlands, which also increases the proportion of mercury that is methylated.

Nitric Acid: Nitric acid presents the potential for acidification of surface water, although there is no known permanent damage in Minnesota. Nitric acid acts as nutrient in nitrogen-poor lakes, such as oligotrophic lakes in northern Minnesota.

Wind Blown Soil

Generally, the size spectrum of wind blown soil particles is sufficiently large that it is not a human health concern for inhalation. However, some components of wind blown soil can have impact on surface water.

Calcium carbonate: Calcium carbonate, a base, neutralizes acid rain in the atmosphere.

Calcium sulfate: Calcium sulfate, which is pH-neutral, can contribute sulfate to sulfate-poor systems, which may stimulate the methylation of mercury.

Phosphorus: Phosphorus is held tightly by soil, so that movement of wind blown soil to surface water can contribute to eutrophication.

Mercury: Soil binds and efficiently holds mercury deposited from the atmosphere, so that the movement of soil to surface water can introduce large amounts of this metal. Lakes in agricultural areas receive high loading of mercury due to soil erosion, but it is unclear whether this mercury is always available for methylation. It is not known how much mercury is carried to lakes by wind blown soil.

Iron: Iron is a limiting nutrient in oligotrophic systems, a phenomenon well documented for the Pacific Ocean and

Lake Tahoe. The oligotrophic lakes in northern Minnesota may also respond to iron additions, although the critical experiments have never been performed. Soil contains significant quantities of iron, so wind blown soil could conceivably fertilize lakes.

Anthropogenic Particulate

Particulate matter is emitted by point sources, area sources, and mobile sources, and often contain materials that might affect surface waters.

Metals: Heavy metals such as cadmium, lead, and silver can be emitted in quantities that are potentially significant to surface water.

Soot: A product of incomplete combustion, soot provides a highly adsorptive surface that can scavenge pollutants such as mercury and dioxin from the atmosphere. Sources of soot include forest fires and poorly tuned combustion devices. Soot may enhance deposition of pollutants to nearby lakes.

Dioxin

Dioxin is a product of incomplete combustion, and also can be formed in processes that utilize chlorine such as paper bleaching. Air emissions of dioxin are extremely low and atmospheric deposition has not been satisfactorily measured. Direct discharge can result in dioxin accumulation in fish in the surface water.

Pesticides

Many pesticides have the potential to cause problems in aquatic systems. Potentially damaging pesticides that have significant deposition rates from the atmosphere include chlordane, DDT/DDE, dieldrin, hexachlorobenzene, alpha-HCH, lindane, and toxaphene. Because of restrictions, none of these currently have significant sources within the United

States. However, volatilization from soils or wind blown soil can deposit significant quantities of these persistent chemicals. In some cases, the compounds are currently used in other countries and transported by the atmosphere to the United States.

Impacts of Chemicals

Many chemicals released by human activity have the potential to disrupt of disrupting the endocrine system of animals, including fish, birds, mammals, and humans. Among these chemicals are persistent, bioaccumulative compounds that include some pesticides, and industrial chemicals such as DDT, lindane, octachlorostyrene, certain PCB congeners, 2,3,7,8-TCDD and other dioxins, 2,3,7,8-TCDF and other furans, atrazine, cadmium, and mercury.

The impacts include thyroid dysfunction in birds and fish, decreased hatching success in birds, fish, and turtles, gross birth deformities, in birds, fish, and turtles, demasculinization and feminization of male fish, birds, and mammals, and defeminization and masculinization of female fish and birds. Many of these compounds are delivered by the atmosphere to aquatic systems.

Ammonia

Like nitrate, atmospheric ammonia that is deposited to lakes and watersheds adds nitrogen to aquatic systems. The addition of nitrogen can contribute to eutrophication, a particular problem in N-limited, oligotrophic lakes in northern Minnesota. Additions of nitrogen may also affect species balances in other systems like prairies and wetlands. The largest sources of ammonia emissions to the atmosphere are: animal agriculture (81%), fertilizer application (10%), refrigeration (5%), and other activities (4%). In terms of total nitrogen emissions to the atmosphere in Minnesota, the major contributors are: animal agriculture

(32%), mobile sources (22%), electric utilities (22%), other fuel combustion (13%), and nitrogen fertilizers (11%).

For most airborne pollutants, it is uncertain what factors might make some geographic regions more sensitive than others. However, it is clear that geological areas low in alkalinity are more sensitive to acid rain. For less obvious reasons, low alkalinity regions are also more sensitive to mercury deposition. These areas of Minnesota are of special concern and will be included in ongoing research into atmospheric deposition of pollutants. Best Management Practices (BMPs) By far the BMP to reduce atmospheric deposition is to halt the release of these pollutants into the atmosphere.

Because of the diversity of sources, cessation of release is complicated and would require the coordination of the full spectrum of the economy, including agriculture, energy production, transportation, waste disposal, manufacturing, and government. Because the atmosphere carries some materials long distances, it may be necessary to address many of these atmospheric pollutants on a national and international basis. Existing BMPs for some other pollutants may lead to some surprising situations.

For instance, it is increasingly common to use wetlands to trap sediments and associated nutrients in storm water before the pollutants can get to a lake or stream. However, the high biological activity of wetlands may lead to some negative consequences for persistent bioaccumulative chemicals. For instance, mercury deposited to terrestrial systems binds strongly to soil particles. Eroded soil may be caught in a wetland, where the mercury would be subject to biological activity.

Because of the heightened activity of anaerobic bacteria that convert sulfate to sulfide, methylation rates are perhaps 100 times higher in wetlands than in lakes. Use of wetlands to clean runoff may therefore enhance methyl mercury loading to surface water, which would increase the concentration of mercury in fish. BMPs for a particular atmospheric pollutant should be selected only after its cycle and fate have been evaluated. Otherwise, we may find ourselves exacerbating the effects of a particular pollutant, as in the hypothetical case of mercury, above.

Another example of the consequences of an incomplete understanding might be attempting to reduce PCBs in Lake Superior by reducing inputs. The PCB burden in Lake Superior is determined by volatilization back to the atmosphere, not external loading. Although research on the environmental fate and budgets of persistent chemicals may be expensive, it is less expensive than making management decisions based on erroneous assumptions, resulting in expensive but ineffective treatment.

3

Waste Water Treatment Systems

Water that has been used by people and is disposed into a receiving water body with altered physical and/or chemical parameters is defined as wastewater. If only the physical parameters of the water were changed, e.g., resulting in an elevated temperature after use as a coolant, treatment before final disposal into a surface water may require only cooling close to its initial temperature. If the water, however, has been contaminated with soluble or insoluble organic or inorganic material, a combination of mechanical, chemical, and/or biological purification procedures may be required to protect the environment from periodic or permanent pollution or damage.

For this reason, legislation in industrialized and in many developing countries has reinforced environmental laws that regulate the maximum allowed residual concentrations of carbon, nitrogen, and phosphorous compounds in purified wastewater, before it is disposed into a river or into any other receiving water body However, enforcement of these laws is not always very strict. Enforcement seems to be related to the economy of the country and thus differs significantly between wealthy industrialized and poor developing countries.

Decomposition of Organic Carbon Compounds

Catabolic processes of microorganisms, algae, yeasts, and lower fungi are the main pathways for total or at least partial mineralization/decomposition of bioorganic and organic compounds in natural or manmade environments. Most of this material is derived directly or indirectly from recent plant or animal biomass. It originates from carbon dioxide fixation via photosynthesis (→ plant biomass), from plants that served as animal feed (→ detritus, feces, urine, etc.), or from fossil fuels or biologi-cally or geochemically transformed biomass (→ peat, coal, oil, natural gas). Even the carbon portion of some xenobiotics can be tracked back to a biological origin, i.e., if these substances were produced from oil, natural gas, or coal.

Only because the mineralization of carbonaceous material from decaying plant and animal biomass in nature under anaerobic conditions with a shortage of water was incomplete, did the formation of fossil oil, natural gas, and coal deposits from biomass occur through biological and/or geochemical transformations. The fossil carbon of natural gas, coal, and oil enters the atmospheric CO_2 cycle again as soon as these compounds are incinerated as fuels or used for energy generation in industry or private households. Biological degradation of recent biomass and of organic chemicals during solid waste or wastewater treatment proceeds either in the presence of molecular oxygen by respiration, under anoxic conditions by denitrification, or under anaerobic conditions by methanogenesis or sulfidogenesis.

Respiration of soluble organic compounds or of extracellularly solubilized biopolymers such as carbohydrates, proteins, fats, or lipids in activated sludge systems leads to the formation of carbon dioxide, water, and a significant amount of surplus sludge. Some ammonia and H_2S may be formed during degradation of sulfur-containing amino acids or heterocyclic compounds. Oxygen must either be supplied by aeration or by injection of pure oxygen. The two process variant for oxygen supply differ mainly in their capacity for oxygen transfer and the stripping efficiency for carbon dioxide from respiration.

Stripping of carbon dioxide is necessary to prevent a drop in pH and to remove heat energy. Respiration in the denitrification process with chemically bound oxygen supplied in the form of nitrate or nitrite abundantly yields dinitrogen. However, some nitrate escapes the reduction to dinitrogen in wastewater treatment plants and contributes about 2% of the total N_2O emissions in Germany. Denitrifiers are aerobic organisms that switch their respiratory metabolism to the utilization of nitrate or nitrite as terminal electron acceptors, if grown under anoxic conditions.

Only if the nitrate in the bulk mass has been used completely does the redox potential become low enough for growth of strictly anaerobic organisms, such as methanogens or sulfate reducers. If anaerobic zones are allowed to form in sludge flocs of an activated sludge system, e.g., by limitation of the oxygen supply, methanogens and sulfate reducers may develop in the center of sludge flocs and form traces of methane and hydrogen sulfide, found in the off-gas.

Under strictly anaerobic conditions, soluble carbon compounds of wastes and wastewater are degraded stepwise to methane, CO_2, NH_3, and H_2S via a syntrophic interaction of fermentative and acetogenic bacteria with methanogens or sulfate reducers. The complete methanogenic degradation of biopolymers or monomers via hydrolysis/fermentation, acetogenesis, and methanogenesis can proceed only at a low H_2 partial pressure, which is maintained mainly by interspecies hydrogen transfer. Interspecies hydrogen transfer is facilitated when acetogens and hydrogenolytic methanogenic bacteria are arranged in proximity in flocs or in a biofilm within short diffusion distances.

The reducing equivalents for carbon dioxide reduction to methane or sulfate reduction to sulfide are derived from the fermentative metabolism, e.g., of clostridia or Eubacterium sp., from β oxidation of fatty acids, or the oxidation of alcohols. Methane and CO_2 are the main products in anaerobic environments where sulfate is absent, but sulfide and CO_2 are the main products if sulfate is present.

Aerobic Biopolymer Degradation

To make soluble and insoluble biopolymers - mainly carbohydrates, proteins, and lipids - accessible for respiration by bacteria, the macromolecules must be hydrolyzed by exoenzymes, which often are produced and excreted only after contact with respective inductors. The exoenzymes adsorb to the biopolymers and hydrolyze them to monomers or at least to oligomers. Only soluble, low molecular weight compounds (e.g., sugars, disaccharides, amino acids, oligopeptides, glycerol, fatty acids) can be taken up by microorganisms and be metabolized for energy production and cell multiplication. Once taken up, degradation via glycolysis (sugars, disaccharides, glycerol), hydrolysis and deamination (amino acids, oligopeptides), or hydrolysis and β oxidation (phospholipids, long-chain fatty acids) proceeds in the cells.

Metabolism of almost all organic compounds leads to the formation of acetyl-CoA as the central intermediate, which is used for biosyntheses, excreted as acetate, or oxidized to CO_2 and reducing equivalents in the tricarboxylic acid (TCA) cycle. The reducing equivalents are respired with molecular oxygen in the respiration chain. The energy of a maximum of only 2 mol of anhydridic phosphate bonds of ATP is conserved during glycolysis of 1 mol of glucose through substrate chain phosphorylation. An additional 2 mol of ATP are formed during oxidation of 2 mol of acetate in the TCA cycle, whereas 34 mol ATP are formed by electron transport chain phosphorylation with oxygen as the terminal electron acceptor.

During oxygen respiration, reducing equivalents react with molecular oxygen in a controlled combustion reaction. When carbohydrates are respired by aerobic bacteria, about one third of the initial energy content is lost as heat, and two thirds are conserved biochemically in 38 phosphoanhydride bonds of ATP. In activated sludge reactors or in wastewater treatment ponds that are not loaded with highly concentrated wastewater, wall irradiation and heat losses with the off-gas stream of aeration into the atmosphere prevent self-heating.

In activated sludge reactors for treatment of highly concentrated wastewater, however, self-heating up to thermophilic temperatures may occur if the wastewater is warm in the beginning, the hydraulic retention time for biological treatment is short (short aeration time), and the air or oxygen stream for aeration is restricted so as to supply just sufficient oxygen for complete oxidation of the pollutants (small aeration volume). The conserved energy in the terminal phosphoanhydride bond of ATP, formed during substrate chain and oxidative phosphorylation by proliferating bacteria is partially used for maintenance metabolism and partially for cell multiplication.

Partitioning between both is not constant, but depends on the nutritional state. In highly loaded activated sludge reactors with a surplus or at least a non-growth-limiting substrate supply, approximately 50% of the substrate is respired in the energy me-tabolism of the cells and 50% serves as a carbon source for cell growth. The biochemically conserved energy must be dissipated to be used for the maintenance metabolism of existing cells and cell growth. If the substrate supply is growth-limiting, e.g., in a low-loaded aerobic treatment system, a higher proportion of ATP is consumed for maintenance, representing the energy proportion that bacteria must spend for non-growth-associated cell survival metabolism, and less energy is available for growth.

Overall, more of the substrate carbon is respired, and the ratio of respiration products to surplus sludge formed is higher, e.g., around 70% : 30%. In a trickling filter system, an even higher proportion of the substrate seems to be respired. This might be due to protozoa grazing off part of the biofilm.

Aerobic Glucose Respiration and Sewage Sludge Stabilization

In most textbooks of microbiology, respiration of organic matter is explained by Eq. 1, with glucose used as a model substance. Except for an exact reaction stoichiometry of the oxidative metabolism, mass and energy dissipation, if mentioned at all, are not quantified. Both parameters are, however, very important for activated sludge treatment plants. The surplus sludge formed during wastewater stabilization requires further treatment, causes disposal costs, and - in the long run - may be an environmental risk, and heat evolution during unevenly high-loaded aerobic treatment may shift the population toward more thermotolerant or thermophilic species and thus, at least for some time, may decrease the process efficiency.

$$1 \text{ mol } C_6H_{12}O_6 + 6 \text{ mol } O_2 \rightarrow 6\text{mol } CO_2 + 6 \text{ mol } H_2O + \text{heat energy} \quad (1)$$

If 1 mol of glucose (MW=180 g) is degraded in an activated sludge system at a high BOD loading rate (e.g., >0.6 kg m^{-3} d^{-1} BOD), approximately 0.5 mol (90 g) is respired to CO_2 and water, with consumption of 3 mol of O_2 (96 g), releasing 19 mol of ATP. The other 0.5 mol of glucose (90 g) is converted to pyruvate by one of three glycolytic pathways, accompanied by the formation of 0.5-1 mol ATP. Pyruvate or its subsequent metabolic products, e.g., acetate or dicarboxylic acids, are directly used as carbon substrates for cell multiplication and surplus biomass formation. A maximum amount of 20 mol ATP is thus available for growth and maintenance. At a pH of 7, about 44 kJ of energy is available for growth per mol of ATP hydrolyzed to ADP and inorganic phosphate.

For an average molar growth yield of aerobes of 4.75 g per mol ATP, 90 g biomass can be generated from 180 g glucose. If the combustion energy per g of cell dry mass is 22 kJ, about 890 kJ (2870-980 kJ) is lost as heat during respiration. The energy loss is the sum of heat losses during respiration and cell growth. At a low BOD loading rate, the proportion of glucose respired in relation to the proportion of glucose fixed as surplus biomass can shift. Up to 0.7 mol (126 g) of glucose can be oxidized to CO_2, requiring 4.2 mol of oxygen (134.4 g O_2). Thus, for respiration of 1 mol of glucose, different amounts of oxygen may be consumed, depending on the loading rate of the wastewater treatment system and the different amounts of carbon dioxide and of surplus sludge formed.

The energy and carbon balance deduced above can be analogously applied to aerobic stabilization of raw sewage sludge. If the initial dry matter content is around 36g L^{-1} (average organic dry matter content of sewage sludge) and if a biodegradability of 50% within the residence time in the sludge reactor is obtained, about 9 g L^{-1} of new biomass is formed, and thus 27 g L^{-1} (36 - 18 + 9) remains in the effluent. The released heat energy is approximately 89 kJ L^{-1} of reactor content. To estimate the theoretical temperature rise, this amount of heat energy must be divided by 4.185 kJ (specific energy requirement for heating 1 L of H_2O from 14.5-15.5 °C).

Thus, by respiration of 18 g L^{-1} organic dry matter, the reactor temperature increases by 21.3°C within the residence time required for degradation (=16 h), provided that no heat energy is lost. A great proportion of the heat energy is, however, transferred via the liquid phase to the aeration gas and stripped out, whereas a smaller proportion is lost through irradiation from the reactor walls. Since air, containing almost 80% nitrogen, is normally used as an oxygen source in aeration ponds or activated sludge reactors, the heat transfer capacity of the off-gas is high enough to prevent a significant increase in the wastewater temperature. Thus, ambient or at least mesophilic temperatures can be maintained.

An increasing temperature of several degrees Celsius would lead to a shift in the population in the reactor and - at least temporarily - would result in reduced process stability, but an only slightly increased temperature of a few degrees Celsius might simply stimulate the metabolic activity of the prevalent mesophilic population. In practice, in activated sewage sludge systems no self-heating is observed because degradability is only about 50% and complete heat transfer to the atmosphere occurs via the off-gas at a retention time of more than 0.5 d. If, however, wastewater from a dairy plant or a brewery with a similar COD concentration, but with almost 100% biodegradable constituents, is stabilized with pure oxygen, twice as much heat evolves, leading to a theoretical temperature rise of 57 °C. Self-heating is observed, since there is much less offgas and the heat loss is thus significantly lower. In addition, due to higher reaction rates than with sewage sludge, the heat is generated during a shorter time span (shorter retention time).

Anaerobic Wastewater or Sludge Treatment

For anaerobic wastewater or sludge treatment, oxygen must be excluded to maintain the low redox potential that is required for survival and metabolic activity of the acetogenic, sulfidogenic, and methanogenic populations. Hydrolysis of polymers, uptake of soluble or solubilized carbon sources, and the primary metabolic reactions of glycolysis up to pyruvate and acetate formation seem to proceed identically or at least analogously in aerobic and anaerobic bacteria. Whereas aerobes oxidize acetate in the TCA cycle and respire the reducing equivalents with oxygen, anaerobes, such as Ruminococcus sp., Clostridium sp., or Eubacterium sp., either release molecular hydrogen or transform pyruvate or acetate to highly reduced metabolites, such as lactate, succinate, ethanol, propionate, or n-butyrate.

For further degradation within the anaerobic food chain, these reduced metabolites must be oxidized anaerobically by acetogenic bacteria. Since the anaerobic oxidation of propionate or n-butyrate by acetogenic bacteria is obligately accompanied by hydrogen production but is only slightly exergonic under conditions of a low H_2 partial pressure, acetogens can grow only when hydrogen is consumed by hydrogen-scavenging organisms such as methanogens or sulfate reducers.

During anaerobic degradation of 1 mol glucose, approximately 95% of the glucose carbon is used for biogas formation (171 g = 127.7 L CH_4 + CO_2), and only about 5% of the substrate carbon (9 g) converted to biomass. Much less heat energy is released during anaerobic metabolism than during aerobic respiration (131 kJ mol^{-1} versus 890 kJ mol^{-1}, respectively), and the biogas contains almost 90% of the energy of the fermented substrate. Due to the heat energy requirement to warm the wastewater and due to heat losses via irradiation from pipes and reactor walls, heat generation is not nearly sufficient to maintain a constant mesophilic fermentation temperature. For this reason, anaerobic digesters must be heated.

In sewage sludge with 36 g L^{-1} organic dry matter content and 50% biodegradability, 0.9 g surplus sludge and 17.1 g biogas (equivalent to 12.75 L) are formed during anaerobic stabilization. Only 13.1 kJ of heat energy per mol of glucose is released, leading to a self-heating potential of 3.1 °C. Since the heat energy is released only during the hydraulic residence time of the wastewater in the reactor, which is usually more than 10 d (except for high-rate industrial wastewater treatment, where in special situations the HRT may be shorter than one day), much more heat energy is lost by irradiation via the reactor walls than is required to maintain the temperature.

If highly concentrated wastewater streams of the food and beverage industry are stabilized anaerobically at hydraulic retention times of <1 d (high space loading), more heat energy is generated within a much shorter time. But even then, process energy from external sources has to be supplied to maintain a temperature of 30-37 °C. Proper insulation of anaerobic reactors can minimize but not replace external heating of the reactor. The methane in the biogas, generated during anaerobic sludge or wastewater stabilization processes, contains about 90% of the energy of the fermented substrate. Since methane is a climate-relevant gas it may not be emitted into the atmosphere but must be combusted to CO_2. Methane from anaerobic reac-tors can be used as a fuel for gas engines to generate electricity and/or heat energy (Eq. 2):

$$CH_4 + 2\ O_2 \rightarrow CO_2 + 2\ H_2O + \text{heat energy}$$
$$(\Delta G^{o\prime} = -891.6\ \text{kJ mol}^{-1}\ CH_4) \qquad (2)$$

Aerobic or Anaerobic Wastewater Treatment Systems

If a producer of wastewater has to decide whether to install an aerobic or an anaerobic waste or wastewater treatment system, several points should be considered: Anaerobic treatment in general does not lead to the low pollution standards of COD, BOD5, or TOC that can be met with aerobic systems and which are required by environmental laws. Anaerobic treatment of wastes and wastewater is often considered a pretreatment process to minimize the oxygen demand and surplus sludge formation in a subsequent aerobic post-treatment stage. Only after a final aerobic treatment can the COD, BOD5, or TOC concentration limits stated in the environmental laws be met.

If limiting concentrations for nitrogen and phosphate also have to be achieved, further treatment steps such as nitrification, denitrification, and biological or chemical phosphate removal, must be considered. Highly concentrated wastewater should in general be treated anaerobically, because of the possibility of energy recovery in biogas and the much lower amounts of surplus sludge to be disposed of. For aerobic treatment, a high aeration rate is necessary and much surplus sludge is generated. Aeration causes aerosol formation and eventually requires off-gas purification. The efficiency of COD degradation for the bulk mass in concentrated wastewater or sludges (degradability of organic pollutants) generally seems to be about similar in aerobic or anaerobic bacteria.

However, the degradation rates may be faster in aerobic treatment procedures than in anaerobic treatment procedures. Wastewater with a low concentration of organic pollutants should be treated aerobically due to its higher process stability at low pollutant concentrations, although aerobic treatment is more expensive and more sludge remains for disposal. If mineralized sludge is required, aerobic treatment at a low loading or at prolonged hydraulic retention times is necessary to reinforce respiration of all endogenous reserve material. Anaerobic treatment systems are more expensive to construct but less expensive to operate than aerobic treatment systems.

Aerobic or Anaerobic Hydrolysis of Biopolymers

Hydrolysis of biopolymers and fermentation or respiration of monomers can be catalyzed by strictly anaerobic, facultative anaerobic, and aerobic microorganisms. With some exceptions (e.g., small protein molecules, dextran), biopolymers are insoluble and form fibers (cellulose), grains (starch) or globules (casein after enzymat-ic precipitation) or can be melted or emulsified (fat). Henze et al. reported hydrolysis constants kh for dissolved organic polymers of 3-20 d^{-1} under aerobic conditions and of 2-20 d^{-1} under anaerobic conditions, whereas for suspended solids the hydrolysis constants kh were 0.6 to 1.4 d^{-1} under aerobic conditions and 0.3 to 0.7 d-1 under anaerobic conditions.

For a kinetic description of hydrolysis and fermentation, a substrate-limited first-order reaction was assumed by Buchauer, who deduced that the temperature-dependent reaction rate for hydrolysis is a little lower than the reaction rate for fermentation of the hydrolysis products. The ratelimiting step is therefore hydrolysis of particles and not fermentation of solubilized material. Since hydrolysis is catalyzed not only by freely soluble exoenzymes, diluted in the bulk mass of liquid, but to a much higher extent by enzymes that are excreted in the neighborhood of bacterial colonies growing attached to the surface of the particles, the above description of complex fermentation processes is not always valid.

Cellulases can be arranged in cellulosomes, which attach to the particles, which in turn serve as carriers until they themselves are solubilized. For this reason, Vavilin et al. included biomass in their description of the hydrolysis of cellulose, cattle manure, and sludge. Shin and Song determined the maximum rates of acidification and methanation for several substrates. For hydrolysis of particulate organic matter, the ratio of surface area to particle size is important. They found that for glucose, starch, carboxymethyl cellulose, casein, and food residues from a restaurant, hydrolysis proceeded faster than methanogenesis, whereas for newspaper and leaves hydrolysis was the rate-limiting step.

Hydrolysis of Cellulose

Cellulose and lignin are the main structural compounds of plants. Both substances are the most abundant biopolymers on earth. Cellulose fibers are formed of linear chains of 100-1400 glucose units linked together by β-1,4-glycosidic bonds. Interand intramolecular hydrogen bonds and van der Waals interactions arrange the highly organized fibrous regions (crystalline region), which alternate with less organized amorphous regions in the cellulose fibers. The fibers are embedded in a matrix of hemicelluloses, pectin, or lignin. The hemicelluloses consist mainly of xylans or glucomannans, which have sidechains of acetyl, gluconuryl, or arabinofuranosyl units. To make cellulose fibers accessible to microorganisms, the hemicellulose, pectin, or lignin matrix must be degraded microbiologically or solubilized chemically.

Cellulose degradation in the presence of oxygen in soil or in the absence of oxygen in the rumen of ruminants, in swamps, or in anaerobic digesters is the most important step in mineralization of decaying plant material. Cellulolytic organisms are found among aerobic soil fungi, e.g., within the genera Trichoderma and Phanaerochaete and in anaerobic rumen fungi, e.g. Neocallimastix and Piromyces, and among bacteria, e.g., within the genera Cellulomonas, Pseudomonas, and Thermomonospora (aerobic cellulose degraders) and Clostridium, Fibrobacter, Bacteroides, and Ruminococcus (anaerobic cellulose degraders).

Glycosyl hydrolases are involved in cellulose and hemicellulose degradation by cleaving glycosidic bonds between different carbohydrates and between carbohydrates and noncarbohydrates. Endo- and exocellulases - in some organisms organized in cellulosomes - must be excreted into the medium. Cellulases are complex biocatalysts and contain a catalytic site and a substrate-binding site. The presence of a noncatalytic substrate binding site permits tight attachment to the different forms of cellulose substrate and keeps the enzyme close to its cleaving sites. Substrate binding is reversible, which allows the enzyme to 'hike' along the fibers and obtain total solubilization.

Many aerobic fungi and some bacteria excrete endoglucanases that hydrolyze the amorphous region of cellulose (degradation within the chain), whereas exoglucanases hydrolyze cellulose from the ends of the glucose chains. Cellobiose is cleaved off by cellobiohydrolases from the nonreducing ends in the amorphous region, and finally the crystalline region is also hydrolyzed. Cleavage of cellobiose to glucose units by β-glucosidases is necessary to prevent cellobiose accumulation, which inhibits cellobiohydrolases. Anaerobic bacteria such as Clostridium thermocellum form a stable enzyme complex, the cellulosome, at the cell surface.

Cellulosomes are active in degrading crystalline cellulose. The catalytic subunits of a cellulosome, endoglucanases and xylanases, cleave the cellulose fiber into fragments, which are simultaneously degraded further by β-glucosidases. Cellulosome-like proteins are found also in Ruminococcus sp. and Fibrobacter sp. cultures. The cell-bound enzymes are associated with the capsule or the outer membrane. Other specific adhesions or ligand formations with the cellulose can be facilitated by fimbrial connections, glycosylated epitopes of carbohydrate binding proteins, or the glycocalyx and carbohydrate binding modules.

Some bacteria have not developed a mechanism to adhere to cellulose fibers, but excrete cellulases into the medium. Adsorption of bacteria onto cellulose fibers via cellulosomes offers the advantage of close contact with the substrate, which is hydrolyzed mainly to glucose, which is then taken up and metabolized. Small amounts of cellobiose must be present initially to induce cellulase expression. The contact of bacteria with the solid substrate surface keeps them close to cellobiose and thus keeps cellulase activities high. However, accumulation of hydrolysis products such as glucose repress cellulase activity.

In nature most cellulose is degraded aerobically. Only about 5%-10% is thought to be degraded anaerobically, which may be an underestimate. Since most ecosystems are rich in carbonaceous substances but deficient in nitrogen compounds, many cellulolytic bacteria can also fix dinitrogen. This is advantageous to them and to syntrophic or symbiotic organisms. Other examples of mutual interactions between organisms of an ecosystem are interspecies hydrogen transfer between anaerobes (anaerobic food chain), transfer of growth factors (mycorhizzae, Kefir), and production of fermentable substrates for the partner organisms (bacterial interactions in the rumen).

Hydrolysis of biological structural components such as cellulose, lignin, and other structural or storage polymers is difficult. The limiting step of hydrol-ysis seems to be liberation of the cleavage products. In contrast to the slow hydrolysis of celluloses, mainly due to lignin encrustation of naturally occurring celluloses, starch can be easily hydrolyzed. The branching and helical structure of starch facilitates hydrolysis.

Whereas cellulose forms fibers with a large surface covered with lignin, starch forms grains with an unfavorable surface-to-volume ratio for enzymatic cleavage. Thus, although amylases may be present in high concentrations, the hydrolysis rate is limited by the limited access of the enzymes to the substrate. Whereas cellulose and starch are biodegradable, other carbohydrate-derived cellular compounds are not biodegradable and - after reaction with proteins - form humic acid-like residues by the Maillard reaction.

Biomass Degradation

In ecosystems in which molecular oxygen is available, plant and animal biomass is degraded to CO_2 and H_2O, catalyzed by either single species of aerobic microorganisms or the whole population of the ecosystem, in competition for the substrates. A single organism may be able to hydrolyze the polymers and oxidize the monomers to CO_2 and H_2O with oxygen. In ecosystems where molecular oxygen is deficient - such as swamps, wet soil, the rumen of animals, the digestive tract of humans, or in river and lake sediments - oxidation of dead biomass proceeds anoxically by reduction of electron acceptors such as nitrate and nitrite or anaerobically by reduction of sulfate, Fe^{3+}, Mn^{4+}, or CO_2.

The oxidation of the carbon source is either complete or incomplete with acetate excretion. In the absence of inorganic electron acceptors, oxidized metabolites such as pyruvate or acetate are reduced to lactate or ethanol or biotransformed to, e.g., n-butyrate or n-butanol. In permanently anaerobic ecosystems with seasonal overfeeding, periodic accumulation of such metabolites can occur, e.g., in autumn after the non-evergreen plants drop their leaves or decay completely. The biopolymers of the leaves or the plants themselves decompose by extracellular enzymatic hydrolysis. The monomers are fermented, and the fermentation products may be degraded further to biogas by acetogenic and methanogenic bacteria.

Whereas single cultures of aerobes can catalyze the whole mineralization process to finally form CO_2 and H_2O, single cultures of strictly anaerobic bacteria are not capable of degrading biopolymers to CH4 and CO_2. Under anaerobic conditions biopolymers must be degraded by a food chain via depolymerization (hydrolysis), fermentation (acidogenesis), oxidation of fatty acids (acetogenesis), and biogas formation (methanogenesis) as the last step. In an initial exoenzyme-catalyzed reaction the biopolymers are hydrolyzed to soluble mono-, di-, or oligomers. These are taken up by the bacteria and fermented to CO_2, H_2, formate, acetate, propionate, butyrate, lactate, etc.

If fatty acid isomers are produced, they are mainly derived from degradation of amino acids after proteolysis. Fatty acids are further oxidized by acetogenic bacteria, before the cleavage products CO_2, H_2, and acetate can be taken up by methanogens and be converted to methane and CO_2. Lactate is oxidized to pyruvate, which is decarboxylated to yield acetate, CO_2, and H_2. If ethanol is present, it is oxidized to acetate and hydrogen, and the hydrogen is used for CO_2 reduction.

In sewage digesters about two thirds of the methane is derived from acetate cleavage and one third from CO_2 reduction with H_2. If hexoses are the substrates and glycolysis is the main degradation pathway, then the 2 mol of pyruvate can be decarboxylated by pyruvate: ferredoxin oxidoreductase to yield 2 mol acetate and 2 mol CO_2. The hydrogens of the 2 mol $NADH_2$ from glycolysis and the 2 mol FdH_2 from pyruvate decarboxylation are then released as molecular hydrogen at low H_2 partial pressure (Eq. 3). Two moles of CH_4 are then formed from acetate and 1 mol of CH_4 by CO_2 reduction.

$$\text{1 mol glucose} \rightarrow \text{2 mol acetate} + \text{2 mol } CO_2 + \text{4 mol } H_2 \text{ (at low } pH_2) \quad (3)$$

In complex ecosystems formate is formed if high concentrations of hydrogen accumulate. Syntrophic interactions are usually associated with interspecies hydrogen transfer, but evidence for interspecies formate transfer was also reported. The feasibility of the electron carrier depends on its solubility, which is much less for hydrogen than for formate, and on its diffusion coefficient in water, which favors hydrogen 30 times over formate. The efficiency of the appropriate electron transfer depends mainly on the distance between the producing and consuming bacteria.

It can be expected, that formate transfer is favored when the distance between communicating bacteria is high and hydrogen transfer when the distance is small. Interspecies formate transfer is thought to play a major role in degradation of syntrophic butyrate and propionate.

However, at an increased H_2 partial pressure formate is also produced by methanogens, either in pure cultures or in a sewage sludge population, and this may also contribute to increasing formate concentrations.

Other substrates for methanogenic bacteria, such as methanol (derived, e.g., from methoxy groups of lignin monomers) or methyl-amines and dimethylsulfide are relevant only in ecosystems where these substances are produced during microbial decay. A few methanogens can also use reduced products such as primary, secondary, and cyclic alcohols as a source of electrons for CO_2 reduction.

Anaerobic Conversion of Biopolymers to Methane and CO_2

Hydrogen seemed to be the most sensitive regulator of anaerobic degradation, formate, acetate, or other fatty acids accumulated to much higher concentrations but did not repress anaerobic degradation. If hydrogen accumulated due to an oversupply or to inhibition of methanogens, anaerobic biodegradation was disturbed successively in different stages. Initially, in the acetogenic stage β oxidation of fatty acids and alcohols failed, leading to an accumulation of these acid metabolites. Later, the spectrum of metabolites of the fermentative flora changed toward more reduced products like ethanol, lactate, propionate, and nbutyrate, leading to an even higher concentration of volatile fatty acids and a further decrease in pH. At pH values below 6.5, methanogenic reactions were almost completely prevented. At this stage acidification with a rich spectrum of reduced products still proceeded.

Anaerobic Degradation of Carbohydrates in Wastewater

Carbohydrates are homo- or heteropolymers of hexoses, pentoses, or sugar derivatives, which occur in soluble form or as particles, forming grains or fibers of various sizes. In some plants, starch forms grains up to 1 mm in diameter, which is 1000 times the diameter of bacteria. Starch metabolism by bacteria requires hydrolytic cleavage by amylases to form soluble monomers or dimers, since only soluble substrates can be taken up and metabolized. The anaerobic degradation of biopolymers in general and of cellulose in particular can be divided into hydrolytic, fermentative, acetogenic, and methanogenic phases.

Hydrolysis and fermentation of the hydrolysis products can be catalyzed by the same trophic group of microorganisms. The distinction of the two phases is of more theoretical than practical relevance. Concerning reaction rates in a methane fermenter that is fed with a particulate substrate, the rate-limiting step is hydrolysis rather than the subsequent fermentation of the monomers, if acetogenesis and methanogenesis proceed faster. The hydrolysis rates of polymers can be very different.

Hemicellulose and pectin are hydrolyzed ten times faster than lignin-encrusted cellulose. In the acidification reactor of a two-stage anaerobic process, hydrolysis of polymers to monomers is normally slower than fermentation of monomers to fatty acids and other fermentation products. For this reason, no sugar monomers can be detected during steady-state operation. In the methane reactor, β oxidation of fatty acids, especially of propionate or n-butyrate, is the ratelimiting step. Fatty acid degradation is the slowest reaction overall in a two-stage methane reactor fed with carbohydrate-containing wastewater from sugar production.

Thus, the methane reactor has to be larger than the acidification reactor to permit longer hydraulic retention times. The rate of cellulose degradation depends strongly on the state of the cellulose in the wastewater. If cellulose is lignin-encrusted, lignin prevents access of cellulases to the cellulose fibers. If cellulose is mainly in a crystalline form, cellulases can easily attach to it, and then hydrolysis can be a relatively fast process. At increasing loading in an anaerobic reactor fed with crystalline cellulose, acetogenesis became the rate-limiting process, leading to propionate and butyrate formation.

In decaying plant material, cellulose is very often lignin-encrusted. Due to the highly restricted access to these complexes by cellulases, hydrolysis of cellulose is the rate-limiting step in its degradation to methane and CO_2. In a natural environment without time limitation, lignin was reported to be degraded anaerobically. However, since these results were based on long-term experiments performed in situ and anaerobiosis was not controlled, it remains doubtful, whether the small amount of lignin that disappeared was really degraded under strictly anaerobic conditions.

The occurrence of coal and fossil oil suggests that lignin compounds are highly resistant to microbial attack. During anaerobic degradation of starch, hydrolysis by amylases proceeds with high velocity if good contact between starch grains and amylases is maintained. Whereas in an anaerobic reactor at low loading, starch degradation can proceed in the absence of acetogenic bacteria at high loading, volatile fatty acids are formed and acetogens are essential for total degradation.

The rationale behind this is that many fermentative bacteria produce only acetate, formate, CO_2, and hydrogen when H_2-scavenging methanogens or sulfate reducers are able to maintain a sufficiently low H_2 partial pressure, but a wide spectrum of fermentation products, typical for the metabolism of the respective bacterial species in pure culture, is produced at higher H_2 partial pressure. Methanogenesis in continuous syntrophic methanogenic cultures can be disturbed by spike concentrations of sugars or - at low concentrations of sugars - by the presence of inhibitory substances like NH_3, H_2S, antibiotics, or xenobiotics. In consequence, the H_2 partial pressure increases and volatile fatty acids are generated. Once propionate or n-butyrate are produced, anaerobic degradation requires acetogens for β oxidation.

Anaerobic Degradation of Protein

Proteins are biological macromolecules, either soluble or solid (e.g., feathers, hair, nails). Outside the cell at an acid pH or in the presence of enzymes, soluble proteins precipitate, e.g., precipitation of casein by addition of rennet enzyme. Hydrolysis of precipitated or soluble protein is catalyzed by sev-eral types of proteases that cleave membrane-permeable amino acids, dipeptides, or oligopeptides. In contrast to the hydrolysis of carbohydrates, which proceeds favorably at a slightly acid pH, optimal hydrolysis of proteins requires a neutral or weakly alkaline pH.

In contrast to the fermentation of carbohydrates, which lowers the pH due to volatile fatty acid formation, fermentation of amino acids in wastewater reactors does not lead to a significant pH change, due to acid and ammonia formation. Acidification of protein-containing wastewater proceeds optimally at pH values of 7 or higher, and ammonium ions together with the CO_2-bicarbonate-carbonate buffer system stabilize the pH. Acetogenesis of fatty acids from deamination of amino acids requires a low H_2 partial pressure for the same reasons as for carbohydrate degradation. This can be maintained by a syntrophic interaction of fermentative, protein-degrading bacteria and acetogenic and methanogenic or sulfate-reducing bacteria.

Except for syntrophic interaction of amino acid-degrading bacteria with methanogens for maintenance of a low H_2 partial pressure, clostridia and presumably also some other sludge bacteria may couple oxidative and reductive amino acid conversion via the Stickland reaction. One amino acid, e.g., alanine (Eq. 4) is oxidatively decarboxylated and the hydrogen or reducing equivalents produced during this reaction are used to reductively convert another amino acid, e.g., glycine, to acetate and ammonia (Eq. 5).

$$CH_3\text{–}CHNH_2\text{–}COOH + 2\,H_2O \rightarrow CH_3\text{–}COOH + CO_2 + NH_3 + 2\,H_2 \quad (4)$$

$\Delta G^{\circ\prime} = +7.5$ kJ mol^{-1}

$$2\,CH_2NH_2\text{-}COOH + 2\,H_2 \rightarrow 2\,CH_3COOH + 2\,NH_3 \quad (5)$$

$\Delta G^{\circ\prime} = -38.9$ kJ mol^{-1}

For complete degradation of amino acids in an anaerobic system therefore, a syntrophism of amino acid-fermenting anaerobic bacteria with methanogens or sulfate reducers is required. If long-chain amino acids are deaminated (Eqs. 6-9), fatty acids such as propionate, i-butyrate, or i-valerate are formed directly. The fatty acids require acetogenic bacteria for their degradation.

$$\text{valine} + 2\,H_2O \rightarrow \text{i-butyrate} + CO_2 + NH_3 + 2\,H_2 \quad (6)$$

$$\text{leucine} + 2\,H_2O \rightarrow \text{i-valerate} + CO_2 + NH_3 + 2\,H_2 \quad (7)$$

$$\text{i-leucine} + 2\,H_2O \rightarrow \text{2-methylbutyrate} + CO_2 + NH_3 + 2\,H_2 \quad (8)$$

$$\text{glutamate} + 2\,H_2O \rightarrow \text{propionate} + 2\,CO_2 + NH_3 + 2\,H_2 \quad (9)$$

In contrast to carbohydrate degradation, where the necessity for propionate- and butyrate-degrading acetogenic bacteria can be circumvented by substrate limitation, during protein degradation these fatty acids are a product of deamination, and their formation cannot be avoided by maintaining a low H_2 partial pressure. In the methanogenic phase there is no difference in methanogenic activity whether carbohydrates or proteins are fermented, except that the methanogens in a reactor fed with protein need to be more tolerant to ammonia and higher pH.

Neutral Fats and Lipids

Fats and lipids are another group of biopolymers that contribute significantly to the COD in sewage sludge, cattle and swine manures, and wastewater from the food industry, e.g., slaughterhouses or potato chip factories. To provide a maximum surface for hydrolytic cleavage by lipases or phospholipases, solid fats, lipids, or oils must be emulsified. Glycerol and saturated and unsaturated fatty acids (palmitic acid, linolic acid, linolenic acid, stearic acid, etc.) are formed from neutral fats. Lipolysis of phospholipids generates fatty acids, glycerol, alcohols (serine, ethanolamine, choline, inositol), and phosphate.

Lipolysis of sphingolipids generates fatty acids and amino alcohols (e.g., sphingosine), and lipolysis of glycolipids generates fatty acids, amino alcohols, and hexoses (glucose, galactose). Sugar moieties and glycerol can be degraded to methane and CO_2 by interaction between fermentative and methanogenic bacteria in low-loaded systems or by cooperation between fermentative, acetogenic, and methanogenic bacteria in high-loaded systems. The long-chain fatty acids are degraded by acetogenic bacteria by β oxidation to acetate and molecular hydrogen.

If acetate and molecular hydrogen accumulate, the anaerobic digestion process is inhibited. Odd-numbered fatty acids are degraded to acetate, propionate, and hydrogen, and even-numbered fatty acids to acetate and hydrogen. Only at a very low H_2 partial pressure, which can be maintained by hydrogen-utilizing methanogens or sulfate reducers, is β oxidation of at least n-butyrate or propionate exergonic. Methanol, ethanol, and ammonia are formed from choline. After hydrolysis, fermentation, and acetogenesis of the fat components in the methanogenic phase acetate, CO_2, and hydrogen are converted to biogas. All subsequent intracellular reactions can be influenced by syntrophic interaction via interspecies hydrogen transfer, except for the extracellular initial lipase reaction.

Carbohydrates, proteins, fats, and biogenic oils can also be degraded anaerobically under thermophilic conditions. The overall degradation scheme is the same, but the microorganism populations are different. Thermophilic fat degradation is becoming more important in practice, since waste fat from fat separators and fat flotates of the food industry are often cofermented in agricultural biogas plants. Since, for hygienic reasons, the input material must be autoclaved, a thermophilic process should be used, keeping the fat in a melted, soluble form for more effective metabolism. Biogas plants with cofermentation of waste fat residues are considered waste treatment systems and must be designed to meet the hygienic demands relevant to treatment of the respective waste.

Methanogens in Methane Reactors

Municipal wastewater or wastewater from sugar production, slaughterhouses, breweries, etc., normally contains less than 200 mg L^{-1} of sulfate. If sulfuric acid is used, e.g., to clean stainless steel containers and pipes in the dairy industry or to maintain an acid pH in bioreactors for bakers' yeast or citric acid production, or if ammonium sulfate is used to inhibit metabolic routes in bakers' yeast for the production of biochemicals, the wastewater contains large amounts of sulfate. Sulfite-containing wastewater is also generated by the starch and cellulose industry during bleaching of the raw products. If these wastewaters are subjected to anaerobic treatment, the methanogenic bacteria must compete with sulfate reducers for the hydrogen equivalents from COD degradation.

In anaerobic digesters sulfate reduction is the favored reaction, due to the higher affinity of sulfate reducers for reducing equivalents or hydrogen, leading to sulfide production at the expense of reduced biogas formation. Sulfide subsequently forms heavy metal precipitates and, if still present in larger amounts, remains in solution and may be toxic to acetogens and methanogens. Some hydrogen sulfide leaves the reactor with the biogas, which may contain up to a few percent of H_2S. Then gas purification is necessary before the gas can be used as fuel for gas engines. Hydrogen sulfide in biogas causes not only odor problems but also corrosion of fermenters and pipes.

Methanogenesis of dissolved and particulate organic material in sulfate-rich wastewater is possible, if acidogenesis and sulfate reduction (stripping of H_2S together with CO_2) is separated from methanogenesis in a phased or a staged process. The higher affinity of sulfate reducers than methanogens for reducing equivalents can also be used for sulfate and heavy metal ion removal from wastewater in the first stage of a two-stage anaerobic process. In comparison to methanogens, which have a rather restricted substrate spectrum, sulfate reducers are metabolically more versatile. Sulfate reducers can utilize polymers such as starch, monomers such as sugars, fatty acids, formate, aliphatic and aromatic compounds, as well as molecular hydrogen to generate reducing equivalents for sulfate reduction (Eq. 10):

$$8\ (H) + 2\ H^+ + SO_4^{2-} \rightarrow H_2S + 4\ H_2O \qquad (10)$$

Undissociated hydrogen sulfide is toxic to both methanogens and sulfate reducers. A 50% inhibition of methanogenesis was observed at a total sulfide concentration of 270 mg L^{-1}, whereas 85 mg L^{-1} sulfide inhibited sulfate reducers. For stable methanogenesis, no more than 150 mg L^{-1} sulfide should be accumulated. H_2S toxicity can be avoided by intensive flushing with biogas to strip the H_2S for adsorption onto $Fe_2(OH)_3$. If large amounts of H_2S are formed during anaerobic treatment, essential heavy metal ions for methanogens (Fe, Ni, Mo, Co, etc.) may be precipitated as sulfides, which may lead to deficiencies in heavy metal bioavailability for the wastewater population.

The metabolites of the fermentative and acetogenia phase of anaerobic wastewater treatment systems, mainly acetate and CO_2 + H_2, are substrates for methanogens and for sulfate reducers. If wastewater with a high sulfate concentration is treated in a methane reactor, the population may gradually shift from hydrogenotrophic methanogens toward hydrogenotrophic sulfate reducers, due to a more favorable Ks value for hydrogen of the sulfate reducers. For pure cultures of Desulfovibrio sp. the Ks value for hydrogen is 1 μM, but for hydrogenotrophic methanogens it is only 6 μM. For acetate, such a dramatic disadvantage of methanogens was not observed. The affinities of sulfate reducers and of methanogenic bacteria (valid at least for Methanosaeta sp.) for hydrogen are in the same range, 0.2-0.4 μM.

Composition of Biogas

From an economic point of view, the specific biogas amounts and the biogas quality from anaerobic treatment of wastewater and sludge are important process parameters. Therefore, a basis for prediction of amount and composition of biogas was elaborated as early as the 1930s. If the elemental composition of a wastewater is known, the theoretical amount and composition of the biogas can be predicted with the Buswell equation (Eq. 11). The actual biogas amount is lower and can be calculated by including a correction factor for the degree of degradability, the pH (which influences CO_2 absorption), and a 5%-10% discount for biomass formation.

$$C_cH_hO_oN_nS_{s+1}/4\ (4c-h-2o+3n+2s)H_2O \rightarrow 1/8\ (4c - h+2o+3n+2s)CO_2 \quad (11)$$
$$\rightarrow 1/8\ (4c + h - 2o - 3n - 2s)CH_4 + nNH3 + sH_2S$$

According to the Buswell equation (Eq. 11), for the anaerobic treatment of a wastewater with carbohydrates as pollutants, the gas composition should theoretically be 50% methane and 50% CO_2 (Eq. 12):

$$1\ mol\ C_6H_{12}O_6 \rightarrow 3\ mol\ CH_4 + 3\ mol\ CO_2 \quad (12)$$

Since CO_2 is increasingly soluble in water with decreasing temperature and increasing pH, CO_2 reacts to form bicarbonate/carbonate, and the biogas may contain more than 80% methane. The total amount of gas is then diminished by the amount of CO_2 that is absorbed and solubilized in the liquid. From a fat- and protein-containing wastewater, theoretically more than 50% methane can be generated.

Removal of Nitrogen During Wastewater Treatment

Nitrogen-containing substances in wastewater are inorganic or organic. Together with phosphates, they represent the main source for eutrophication of surface water. For this reason they must be eliminated together with the organic carbon during wastewater treatment. Whereas phosphates form insoluble precipitates with many heavy metal ions and the precipitates can be separated by sedimentation or flotation, all nitrogen compounds, except for magnesium ammonium phosphate, are easily soluble in water and thus cannot be removed chemically by precipitation.

For biological removal of amino nitrogen and of heterocyclic nitrogen compounds, their conversion to ammonia in an aerobic or anaerobic treatment process is the first step. Then ammonia must be nitrified, and the nitrate denitrified to yield nitrogen. Thus, depending on the kind of nitrogen compounds present in wastewater, nitrogen removal requires up to three processes in sequence: ammonification, nitrification, and denitrification.

The major portion of nitrogen compounds in municipal wastewater are reduced nitrogen compounds such as ammonia, urea, amines, amino acids, and proteins. Oxidized nitrogen compounds such as nitrate and nitrite normally are not present at all or not in relevant amounts. Nitrate and nitrite may, however, represent the main nitrogen load in wastewater of certain food or metal industry branches.

The ammonia in raw municipal wastewater is mainly derived from urine and is formed in the sewer system by enzymatic cleavage of urea by ureases (Eq. 13):

$$NH_2CONH_2 + H_2O \rightarrow CO_2 + 2NH_3 \qquad (13)$$

The residence time of the wastewater in the sewer system normally is not long enough for a significant contribution of ammonia from other sources, e.g., proteolysis and deamination of amino acids.

Ammonification

The main organic nitrogen compounds in municipal wastewater are heterocyclic compounds (e.g., nucleic acids) and proteins. Proteolysis and degradation of amino acids leads to liberation of ammonia by the various mechanisms of ammonification, including hydrolytic, oxidative, reductive, and desaturative deamination (Eqs. 14-17, respectively)

$$R\text{-}NH_2 + H_2O \rightarrow R\text{-}OH + NH_3 \qquad (14)$$

$$R\text{-}CHNH_2COOH + H_2O \rightarrow R\text{-}CO\text{-}COOH + 2\,(H) + NH_3 \qquad (15)$$

$$R - CHNH_2 - COOH + 2\ (H) \rightarrow R - CH_2 - COOH + NH_3 \quad (16)$$

$$R\text{–}CH_2\text{–}CHNH_2\text{–}COOH \rightarrow R\text{–}CH{=}CH\text{–}COOH + NH_3 \quad (17)$$

A significant amount of ammonia from urea cleavage or from ammonification of amino acids is assimilated in aerobic treatment processes for growth of bacteria (surplus sludge formation). It can be estimated that bacteria consist of roughly 50% protein and that the nitrogen content of protein is about 16%. Thus, for synthesis of 1 g of bacterial biomass, about 0.08 g of ammonia-N is required. To eliminate ammonia that is not used for cell growth during wastewater treatment, it must first be nitrified and then denitrified to molecular nitrogen or anaerobically oxidized with nitrite.

Nitrification of Ammonia

Autotrophic nitrification

Autotrophic nitrifiers are aerobic microorganisms oxidizing ammonia via nitrite (Eq. 18) to nitrate (Eq. 19). Organisms catalyzing nitrification (Eq. 18) belong to the genera Nitrosomonas, Nitrosococcus, Nitrosolobus, Nitrosospira, and Nitrosovibrio, organisms catalyzing nitration (Eq. 19) include, e.g., members of the genera Nitrobacter, Nitrococcus, and Nitrospira.

$$NH_4^+ + 1.5\ O_2 \rightarrow NO_2^- + 2\ H^+ + H_2O \quad (18)$$

$$NO_2^- + 0.5\ O_2 \rightarrow NO_3^- \quad (19)$$

Ammonia oxidation to nitrite or nitrite oxidation to nitrate are energy-yielding processes for the autotrophic growth of the nitrifying bacteria. CO_2 is assimilated via the Calvin cycle. Since the positive redox potential of the oxidizable nitrogen compounds is not low enough to form $NADH_2$ for CO_2 reduction, $NADH_2$ must be formed by an energy-consuming reverse electron transport. For this reason, the growth yield of nitrifying bacteria is low. Nitrosomonas sp., e.g., must oxidize 30 g NH_3 to form 1 g of cell dry mass. The oxidation of ammonia by nitrifiers is initiated by an energy-neutral monooxygenase reaction that yields hydroxylamine (Eq. 20):

$$NH_3 + XH_2 + O_2 \rightarrow NH_2OH + X + H_2O \qquad (20)$$

Then hydroxylamine is further oxidized, presumably via nitroxyl (Eq. 21) to nitrite (Eq. 22), which is the energy-yielding reaction during nitrification. For microbial oxidation of 1 mg NH_4^+-nitrogen to nitrite, 3.42 mg O_2 is required, and for the oxidation of 1 mg NO_2^--nitrogen to nitrate (Eq. 19), 1.14 mg O_2.

$$NH_2OH + X \rightarrow XH_2 + (NOH) \qquad (21)$$

$$(NOH) + 0.5\ O_2 \rightarrow HNO_2 + \text{energy} \qquad (22)$$

Due to their slow growth autotrophic nitrifiers cannot successfully compete with heterotrophic bacteria for oxygen. In a highly loaded activated sludge system, the autotrophic nitrifiers are overgrown by the heterotrophic sludge flora, which consume the oxygen. Ammonia oxidation starts only if the BOD_5 concentration in the wastewater is <110 mg L^{-1}. During nitrification of ammonia the alkalinity of the wastewater increases slightly, due to CO_2 consumption for autotrophic growth (pH increase), but in a counter reaction it drastically decreases due to nitric acid formation from ammonia (pH decrease from above neutral to acidic). If the buffer capacity of the wastewater is weak, the pH drops far below 7 and thus prevents further nitrification by autotrophic nitrifiers.

Heterotrophic nitrification

Some bacteria of the genera Arthrobacter, Flavobacterium, and Thiosphaera are able to catalyze heterotrophic nitrification of nitrogen-containing organic substances (Eq. 23):

$$R\text{-}NH_2 \rightarrow R\text{-}NHOH \rightarrow R\text{-}NO \rightarrow NO_3^- \qquad (23)$$

Heterotrophic nitrifiers oxidize reduced nitrogen compounds, such as hydroxylamine and aliphatic and aromatic nitrogen-containing compounds, but in contrast to autotrophic nitrification, no energy is gained by nitrate formation. For this reason an organic substrate must be respired to satisfy the energy metabolism. Some heterotrophic nitrifiers can denitrify nitrate or nitrite under aerobic growth conditions.

The nitrogen metabolism of Thiosphaera pantotropha and Paracoccus denitrificans are well documented examples. These organisms express a membrane-bound nitrate reductase under anoxic growth conditions, which works only in the absence of molecular oxygen. Under aerobic growth conditions a periplasmic nitrate reductase is expressed, which catalyzes nitrate reduction at least to the state of nitrous oxide.

Denitrification

Many aerobic bacteria seem to be able to switch their oxidative metabolism to nitrate respiration. Similar to oxygen respiration, the nitrate respiration of heterotrophic bacteria requires a complex carbon source as an electron source for denitrification (e.g., Eqs. 24 and 25). Denitrification (Eq. 26) starts with the reduction of nitrate to nitrite by membrane-bound nitrate reductase A (a). Then a membrane-bound nitrite reductase (b) catalyzes NO formation. Finally, NO reductase (c) and N_2O reductase (d) form N_2.

The theoretical stoichiometry of denitrification with methanol or acetate as a carbon source is shown in Eqs. 27 and 28. For practical application a surplus of carbon source must be supplied, since the wastewater is not free of oxygen, and part of the carbon source is respired until anoxic conditions are achieved. To evaluate the stoichiometry of nitrate to organic compounds for denitrification with a complex carbon source, the oxidation-reduction state of the carbon substrates and the oxygen concentration in the wastewater should be known.

In wastewater treatment plants more than 2.85 g COD is required for reduction of 1 g NO_3^--N.

$$CH_3OH + H_2O \rightarrow CO_2 + 6\ H+ + 6\ e^- \tag{24}$$

$$CH_3COOH + 2\ H_2O \rightarrow 2\ CO_2 + 8\ H^+ + 8\ e^- \tag{25}$$

$$NO_3 \underset{a}{\overset{2e^-}{\rightarrow}} NO_2 \underset{b}{\overset{e^-}{\rightarrow}} NO \underset{c}{\overset{e^-}{\rightarrow}} 0.5\ N_2O \underset{d}{\overset{e^-}{\rightarrow}} 0.5\ N_2 \tag{26}$$

$$5\ CH_3OH + 6\ HNO_3 \rightarrow 5\ CO_2 + 3\ N_2 + 13\ H_2O \tag{27}$$

$$5\ CH_3\text{-}COOH + 8\ HNO_3 \rightarrow 10\ CO_2 + 4\ N_2 + 14\ H_2O \tag{28}$$

Instead of nitrate, many denitrifying bacteria can use NO_2^-, NO, or N_2O as terminal electron acceptors. Alternatively, they may release these intermediates during denitrification of nitrate under unfavorable conditions, as was observed in soil. If, e.g., surplus nitrate is supplied and hydrogen donors are not sufficiently available, NO and N_2O can be formed. Another condition for N_2O formation is a pH below 7.3, at which nitrogen oxidoreductase is inhibited. Except for dissimilatory nitrate reduction, many aerobic and anaerobic bacteria are capable of assimilatory nitrate reduction to supply the cells with ammonia for growth (Eqs. 29 and 30). However, the enzymes for nitrate assimilation are expressed only at concentrations of ammonia <1 mM.

$$NO_3 \underset{a}{\overset{2e^-}{\rightarrow}} NO_2 \underset{b}{\overset{2e^-}{\rightarrow}} HNO \underset{c}{\overset{2e^-}{\rightarrow}} NH_2OH \underset{d}{\overset{2e^-}{\rightarrow}} NH_3 \quad (29)$$

$$HNO_3 + 8\ (H) \rightarrow NH_4 + OH^- + 2\ H_2O \quad (30)$$

Here, nitrate reductase B (enzyme a in Eq. 29) reduces nitrate to nitrite, which is then reduced to ammonia by the nitrite reductase complex (enzyme b in Eq. 29). Whereas nitrate reduction by the oxygen-sensitive, membrane-bound enzyme nitrate reductase A conserves energy, no energy conservation is possible in the reaction catalyzed by the soluble enzyme nitrate reductase B, which is not repressed by oxygen. For details on the cell biology and the molecular basis of denitrification, please refer to Zumft.

New N-removal Processes

The metabolic versatility of nitrogen-converting bacteria offers new concepts for nitrogen removal processes to treat wastewater. A partial nitrification of ammonia to nitrite (nitritation) would have the advantage of saving a significant amount of oxygen for the nitrification process, as compared to nitratation of ammonia. However, nitrite is toxic and must be kept below toxic concentrations. In a subsequent denitrification process, fewer reducing equivalents from the degradation of different carbon sources, which often are a limiting factor in wastewater treatment plants, are then required.

In the SHARON process, which is a single reactor system for rapid ammonia oxidation to nitrite and subsequent denitrification of the nitrite with methanol, nitrite oxidizers that would generate nitrate are outcompeted by the slowly metabilizing ammonia oxidizers after a temperature shift toward higher temperatures, due to the higher temperature tolerance of the ammonia oxidizers, and by no sludge retention or sludge return in the reactor. Aeration was switched off periodically and denitrification of nitrite was initiated by adding methanol as a carbon source. The SHARON process in full scale was used to treat sludge liquor at the Rotterdam wastewater treatment plant.

The partial nitrification of ammonia to nitrite led to an ammonia/nitrite ratio that was suitable for a subsequent Anammox process in which nitrite was denitrified to N_2 with ammonia instead of methanol as the electron donor. This process was finally established in full-scale at the Rotterdam wastewater treatment plant. Partial ammonia oxidation to nitrite and anaerobic oxidation of ammonia in one aerated reactor is the concept behind the CANON process (completely autotrophic nitrogen removal over nitrite). Aerobic and anaerobic ammonia-oxidizing bacteria cooperate as long as the nitrifying bacteria consume oxygen and create anoxic conditions for the anaerobic ammonia-oxidizing bacteria.

Another possibility for removing ammonia in one single step without using an organic carbon source is the OLAND process (oxygen-limited nitrification and denitrification). Conversion of ammonia to N_2 is catalyzed by aerobic nitrifiers and anaerobic granular sludge as a source of Planctomycetes including anaerobic ammonia-oxidizing bacteria. Other nitrogen-removal processes also exist, such as the NOx process (simultaneous nitrification and denitrification of Nitrosomonas-like microorganisms under fully oxic conditions) and aerobic deammonification (conversion of ammonia to N_2 and nitrate), which allow ammonia removal without COD.

Biological Removal, Biotransformation, and Biosorption of Metal Ions

Whereas solid organic and inorganic material in wastewater or sludge can be removed by sedimentation, soluble organic pollutants and xenobiotics should be eliminated from the aqueous environment by microbial mineralization or anaerobic degradation to gaseous products, with a varying portion (5%-50%) being used as substrates for bacterial growth. Most of the inorganic components present in wastewater are soluble and are ionized. Trace amounts of many cations (e.g., Na^+, K^+,Ca^{2+}etc.) and anions (e.g., PO_4^{3-}, Cl^-, S^{2-}) are essential micronutrients for bacterial growth. Other cations such as ammonia may also be required for bacterial growth, but the surplus amount must be oxidized to nitrate, and the nitrate denitrified to gaseous nitrogen, for N elimination from wastewater.

Under anaerobic conditions sulfate is reduced to sulfide, low amounts of which are required for growth of bacteria. The sulfide not required for growth is toxic for bacteria if present as H_2S in high concentrations. In anaerobic reactors at a slightly alkaline pH, most of the sulfide is precipitated as heavy metal sulfides; then it is harmless to microorganisms and their environment. To support detoxification, heavy metal ions may also be precipitated chemically. Metal ion contaminants in wastewater can be removed by microorganisms by either a direct or indirect influence on the redox state of the metal ions or through biosorption of metal ions on the cell surface.

Some microorganisms have also developed resistance mechanisms against toxic metals by changing the oxidation state without supporting anaerobic growth. Certain bacteria, yeasts, fungi, and algae can actively accumulate intracellular metal ions against a gradient. The process of bioaccumulation of metal ions depends on living, metabolically active cells, whereas biosorption is a passive, energy-independent process that can be mediated also by inactive cell material. Biosorption of metal ions includes mechanisms such as ion exchange, chelation, matrix entrapment, and surface sorption. After biosorption or active removal of metal ions from wastewater or contaminated soil, the heavy metal-containing biomass must be separated and incinerated or regenerated by desorption or remobilization of the metals.

As an example of successful biosorption, wastewater from the galvanizing industry that contained 29 mg L^{-1} Zn and 10.5 mg L^{-1} Fe was purified in a threestage stirred-reactor cascade. With 15 g biomass, 7.5 L of wastewater was decontaminated. The removal of metal ions by biologically catalyzed changes in the redox state is an alternative. The altered speciation (valence status) of metals can lead to precipitation, solubilization, or volatilization of the metal ions. Many bacteria use metal ions as electron acceptors for anaerobic respiration. Examples are the reduction of Fe^{3+}, Cr^{6+}, Mn^{4+}, Se^{6+}, As^{5+}, Hg^{2+}, Pb^{2+}, or U^{6+}.

The dissimilatory metal-reducing bacteria can use H_2 or organic pollutants (e.g., xenobiotics) as electron donors and are capable of simultaneously removing organic and inorganic contaminants. Metal ions are reduced and precipitated by sulfide that is generated by sulfate-reducing bacteria. Solubilization of most heavy metal precipitates is favored at acid pH, which is the favorable pH for soil or sludge decontamination, whereas an alkaline pH is favorable for precipitation of heavy metal ions to decontaminate wastewater. An example of the formation of precipitates or soluble compounds at different redox states occur in ferrous and manganese compounds.

Under aerobic conditions Fe(III) (ferric iron) and Mn(IV) ions form insoluble Fe(III) and Mn(IV) oxides or hydroxides, but under anaerobic conditions they form soluble Fe(II) compounds (ferrous iron) or Mn(II) compounds. Most organisms that grow with energy conserved during reduction of Fe(III) or Mn(IV) are members of the Geobacteriaceae. Transfer of electrons from the terminal reductase, localized in the outer membrane or at the cell surface, to the insoluble Fe(III) or Mn(IV) oxides outside the cells can proceed either in a direct way (contact between the oxides and the cells) or by 'soluble electron shuttles' (e.g., by humic substances) between the metal-reducing microorganism and the mineral.

Since Fe is an essential element for microorganisms, aerobic bacteria must excrete siderophores, which bind Fe^{3+} to their phenolate or hydroxamate moiety and supply the cells with soluble Fe^{2+}. To accelerate Fe^{3+} reduction in biotechnological processes, the chelator nitrilotriacetic acid can be added. In addition to their use in synthesis of cell components (e.g., cytochromes, ferredoxin, etc.), Fe^{2+} salts can be electron donors for nitrate reduction. The reduction of Hg^{2+} to metallic Hg by Escherichia coli or Thiobacillus ferrooxidans facilitates Hg separation and prevents methylation reactions under aerobic or sulfate-reducing conditions.

The mechanism includes nonenzymatic transfer of methyl groups from methylcobalamin to Hg^{2+} to form methyl mercury or dimethyl mercury, which are both neurotoxins and become enriched in the food chain. Selenium and arsenic can be transformed microbiologically by methylation to dimethyl selenide or to di- or trimethyl arsine in a volatile form. Methylated arsenic compounds are less toxic than nonmethylated arsenic compounds. Changing the redox state of natural and anthropogenic radionuclides by metal-reducing microorganisms offers a possibility to control their solubility and mobility by converting, e.g., U^{6+} to U^{4+}, Pu^{5+} to Pu^{4+}, or Np^{5+} to Np^{4+}.

The tetravalent metals can be removed by chelators (e.g., EDTA) or by immobilization onto biomass from the contaminated environment. In contrast to bacteria, fungi are capable of leaching soluble as well as insoluble metal salts, because they excrete organic acids such as citric acid, fumaric acid, lactic acid, gluconic acid, oxalic acid, or malic acid, which dissolve metal salts and form complexes with the metal ions.

The leaching efficiency depends on the soil microflora. Some soil microorganisms seem to be able to degrade the carbon skeleton of the metal-organic complex and thus immobilize the metal ions again. Mixed bacterial cultures or Wolinella succinogenes use perchlorate or chlorate as electron acceptors for respiration and thus detoxify these chemicals.

Sulfate Reduction

Sulfate-reducing bacteria are biotechnologically relevant to sulfate removal or heavy metal precipitation in wastewater or waste and to the elimination of SO_2 during offgas purification. An overview of applications of sulfate-reducing microorganisms in environmental biotechnology is given by Lens et al.. Sulfate is the terminal electron acceptor and is reduced to sulfide, with reducing equivalents derived from the degradation of lactic acid or many other organic compounds. Alternatively, some sulfate reducers can also use molecular hydrogen. Sulfate reducers gain energy in an anaerobic electron transport chain, leading to sulfide, a weak dibasic acid, which dissociates according to Eq. 31.

$$H_2S \overset{K_1}{\leftrightarrow} H^+ + HS^- \overset{K_2}{\leftrightarrow} 2\,H^+ + S^{2-}$$
$$K_1 = 1.02 \times 10^{-7},\ K_2 = 1.3 \times 10^{-13}\ (25\ °C) \qquad (31)$$

The total dissociation is described by Eq. 32:

$$K = \frac{[H^+][S^{2+}]}{[H_2S]} = K_1K_2 = 1.3 \times 10^{-20} \qquad (32)$$

For precipitation of heavy metal ions, sulfide ions are necessary (Eq. 33):

$$Me^{2+} + S^{2-} \leftrightarrow MeS \text{ (or) } 2\,Me^+ + S^{2-} \leftrightarrow Me_2S \qquad (33)$$

The concentration of sulfide is pH-dependent. At acid pH only those metal sulfides of very low solubility can be precipitated. Thus, at acid pH, HgS, As_2S_3, CdS, CuS, and PbS form precipitates, whereas at a more alkaline pH, ZnS, FeS, NiS, and MnS form precipitates. Al_2S_3 and Cr_2S_3 are water soluble and cannot be removed by precipitation or sedimentation. Zinc removal from zinc-contaminated groundwater by microbial sulfate reduction and zinc sulfide precipitation in a 9-m^3 sludge blanket reactor was demonstrated and has been transferred to a full scale reactor of 1800 m^3.

4

Solid Waste Management

Decentralised Community Waste Management Systems

Any community project needs the support of its stakeholders - waste pickers, residents, local municipal body, community-based organisations (CBOs), volunteers, etc - to operate and sustain itself. However the terms of such interface or interaction have never been fully examined or adequately documented. The possibilities of upscaling such communitybased projects - either as larger projects or through greater number of such projects - have never been explored.

The current rules and regulations of the government are inadequate to assess the environmental benefits or the economic and social implications of waste projects. There is a strong case for intervention in both upscaling the community projects as well as in making suitable policy interventions to ensure that they are economically and environmentally sustainable. The traditional system of recycling waste, in which the waste pickers play a crucial role, is being mer- cilessly torn apart and displaced by corporatising waste.

As the international waste-to-energy industry gains a foothold in this area, it will burn the future of many urban poor. At the same time, municipalities that appear to support communitybased projects have been quietly signing off waste projects to private interests. For developing countries like India, recycling of waste is the most economical and socially viable option as it generates employment for the urban poor who neither have skills nor capital for investment. The initiative and incentive for community waste management systems should ideally come from the government.

Land for composting and other basic infrastructure should be provided by the local government. Currently, urban planning does not include such needs in its spatial city plans. Markets need to be developed for compost products made from urban waste. Private sector investment in this area has been floundering owing to its inability to sell the compost due to the competition from the heavily subsidised chemical fertiliser industry. While energy products are being subsidised, the greener compost products need urgent attention.

Organised waste collection through community waste management systems should include waste pickers. This will lead to more, and cleaner, recovery of recyclables and enhanced income for the waste pickers. The burgeoning waste situation clearly needs a new framework. Decentralised initiatives carried out by numerous groups in numerous ways - in small local pockets as well as larger colony and zonal levels - are showing the direction for the new paradigm. These need to be negotiated into the system of urban planning, municipal partnerships as well as marketing links.

Community-based Solid Waste Management Initiatives

Basically, there are five types of waste management models in the Decentralised Solid Waste Management Systems (DSWMS):

— *Partnership between Municipality and NGOs*: In this scenario NGOs/CBOs manage community interventions with the assistance of municipalities. For example, Exnora Green Cross (EGC), Vellore and Jan Chaitnya in Vishakhapattanam have been sustaining their intervention with the help of the local municipality.

— *NGOs/CBOs on their own*: The Muskan Jyoti Samiti (MJS)intervention in Lucknow is a good example of this approach. MJS has been given support by State Urban Development Agency (SUDA), Lucknow District Collector and other government agencies, but never by Lucknow Nagar Nigam, which is constitutionally obliged to provide these services.

— *Municipalities on their own*: Municipalities of cities such as Suryapet, Bangalore, Panjim, Kalyani, Bhadreshwar, etc, are engaged in sustainable management of urban solid wastes purely on their own and as per the MSW Rules, 2000.

— *Private Operators*: In some places (Chennai, Nasik, Surat and Delhi), municipalities have contracted out their solid waste management functions to private operators.

— *Institutions/industrial complexes on their own*: Some institutions/industrial complexes manage their solid waste problems with the help of local NGOs. Some examples are Indian Institute for Technology (IIT) and Jawaharlal Nehru University (JNU) in New Delhi and ITC, Bhadrachalam which are managing their solid wastes with the help of local NGOs.

The following discussion pertains to those DSWMS which have been initiated by the NGOs/CBOs and are being run with some support from municipalities or government agencies. From the municipality they receive support which could be in the form of land allocation for segregation/ composting, the issuance of identity kits to waste collectors, etc. These interventions however collect a 'user fee' from the community. The interventions that fall into this category are depicted in the table on the next page. Of these, all except the intervention of Jan Seva Ashram (JSA), at Solan, have up-scaled and are sustaining themselves. The JSA intervention has failed due to various reasons.

The up-scalability and sustainability of these community interventions, are discussed in detail in the next section. Here we can discuss some significant findings and observations about these projects based on the data collected during the process:

Community Projects

A marginal but important presence at the city level: All these community interventions have a very marginal presence. Even if we take the two largest interventions, namely MJS at Lucknow and Centre for Development Communication (CDC) at Jaipur, they deal with only 3.6% and 2.5% of the city waste respectively. However, they make a significant contribution in the form of providing employment to waste collectors. MJS, CDC and Kagach Kach Patara Kashtkari Panchayat (KKPKP) employ 900, 600 and 300 waste collectors respectively.

Kinds of services: Most of these organisations mainly adopt door-to-door collection, segregation and composting of bio-degradable waste. Thus, NGOs like Naya Savera, Centre for Environment and Education (CEE), Sukuki Exnora, Friends of Urban Poor and Vikas are involved in doorto- door collection of waste and other waste management services. Some NGOs like Vatavaran and Stree Mukti Sangathana (SMS) are also involved in staircase cleaning, tree pruning, etc. However, some organisations like Nav Bharat Jagriti Kendra (NBJK), EGC have also integrated drainage cleaning and road sweeping, though not on a daily basis.

Composting: Out of 16 projects, 11 have undertaken composting. Only MJS, Lucknow; NBJK, Ranchi; KKPKP, Pune are not involved in composting. MJS had to stop its composting initiative due to marketing problems while NBJK has a plan to start composting once the municipality allocates land for it. The table below depicts the allocation of land for composting. Municipalities have provided land in half the cases which is quite logical since it is the main agency responsible for the solid waste management of the city.

In some residential societies - as in Basera Colony, Mumbai - RWAs have provided land for local composting. In the case of Naya Savera and Vatavaran, educational institutions like IIT and JNU have provided the land for the composting.

Landfill diversion rate: One of the most important benefits accruing from these community interventions is their ability to divert waste from landfills. The landfill diversion rate acquires significance when one considers the nature of Indian waste. Urban waste in India usually comprises 60% bio-degradable waste, about 30% recyclables and only 10% is absolute waste. Since about 60% of the municipal waste is organic in composition it can easily be composted. Assuming that there is primary segregation taking place in order to ensure the quality of compost, there is an automatic diversion of 60% of the waste.

Moreover, since source segregation ensures separate collection of recyclables and other waste, the rate of landfill diversion goes up further. This is evident from the data collected form Jan Chaithanya's Vishakhapattanam intervention. Jan Chaithanya has been able to achieve a landfill diversion rate of 91.6% through composting of bio-degradable waste and sale of recyclables. Since, 11 out of the 16 surveyed organisations are engaged in composting, it can be safely inferred that they are able to achieve a landfill diversion rate of more than 80%.

Community Interventions Upscaling

All the ongoing interventions, except JSA, Solan, have scaled up their activities with varying degrees. The upscaling has occurred at different levels:

— Upscaling in terms of the increase in the coverage population: All the documented organisations have succeeded in increasing their coverage; but to different extents. This has happened through replication of the prototypes at different residential colonies in the same city. This has been the most frequent type of upscaling and all the interventions have succeeded in doing so.

— Upscaling in terms of different kinds of interventions: Some organisations have been innovative in their approach and have not only upscaled but have also started different kinds of interventions. Exnora Green Cross has been spectacularly successful with its interventions in diverse kinds of settings - at the Vellore Institute of Technology, the local fish market, temples, cow sheds, etc.

— Upscaling in terms of 'replication' in many cities: CDC is only organisation that has been able to replicate its interventions in many cities. Its interventions have spread to Jaipur, Nanded, Nagpur and Surat. However, the nature of the intervention is different at different places.

Community Interventions Sustainability

Based on the data gathered during this documentation process, this section seeks to identify some important basic indicators that determine the sustainability and up-scalability of decentralised community interventions in solid waste management. The period of operation is perhaps the most flexible variable in the sustainability of these interventions. There is no fixed time period which can be ascribed to a sustainable intervention. Surprising as it may sound, it appears that a timescale, though a necessary parameter, is not a sufficient condition to define sustainability.

Ecological sustainability

Ideally, in order to be ecologically sustainable, a community intervention should strive for the following:

— Minimum production of waste.

— Reuse and recycling of waste to a maximum.

— Bio-degradable waste to be treated separately and only inerts to be dumped in landfills.

To fulfill the last two conditions optimally, waste should be segregated at source. However, not all these conditions are being fulfilled by the documented interventions. It is difficult to measure the relevance of waste minimisation since none of the organisations have records showing that they have achieved some degree of waste minimisation since the intervention started. It is the same in the case of reuse and maximum recycling of waste. Since the fulfillment of these two conditions requires behavioural changes - which is a long term process - we need to give some more time to these interventions.

Source segregation of waste is the minimal condition to treat bio-degradable waste, and to maximize recycling recovery. Even that is not being met at every place. All these interventions have collectively achieved only 30% primar y segregation. Two of the most up-scaled interventions - MJS at Lucknow and CDC at Jaipur - have been managing without primary segregation of waste. They have instead opted for on-rickshaw segregation, which can not be considered source segregation. Considering the experience they have under their belts, they can be sited as successful examples. Separate treatment of bio-degradable waste through composting is being done at 11 out of the total 16 interventions.

Thus, it is also not a necessary condition. Though it is definitely an extra source of income and increases the financially sustainability of the project. Thus, in the present urban Indian scenario, primary segregation of waste, composting or separate treatment of bio-degradable waste are neither necessary nor sufficient conditions for sustainability of community interventions in solid waste management. However, there is no denying the importance of these conditions for long term ecological sustainability. In other words this means that:

- There is either enough margin in running these operations or there is significant reduction in establishment costs if the intervening organisation does not undertake composting of biodegradable waste.
- The quality of services being provided by municipalities are so low that the community is ready to pay organisations that are doing just primary collection, without promoting source segregation and composting of bio-degradable.

Financial sustainability

Perhaps, this is the most important aspect of sustainability. It must be recalled that we are discussing only those projects, which are sustaining due to a contribution from the community. This contribution is in the form of a user fee, in lieu of services provided by these organisations. If satisfactory services are provided, the community will pay for them. With 91% consumers feeling comfortable in paying the user fee and 85% feeling that they are being reasonably charged, it can be safely assumed that there is 'willingness' among the community to pay for it. Some of the important issues which should be considered for attaining financial sustainability are:

- Organisations should be pragmatic in their approach over the issue of the user fee. There should be different user fee slabs for different income categories.

- — A significant 39% of the community was either undecided or unwilling to pay more for the services. Hence, organisations should proceed cautiously and should approach the community with sufficient reasons, if there is a need to increase the user fee.
- — Moreover, extras sources of revenues, from the sale of compost and recyclables, should also be considered. Here, the trend is mixed. Some organisations have rights over recyclables and treat it as a source of revenue. Thus, NGOs like Naya Savera and Sukuki Exnora earn about Rupees 15,000 to 20,000 per month from the sale of recyclables and use it to pay the salary of waste collectors. On the other hand, organisations like CDC exercise no right over recyclables and waste collectors are allowed to keep the proceeds from the sale of recyclables.

  Similarly, organisations like Exnora Green Cross and Vikas earn about Rupees 16,000 and 8,000 per month respectively from the sale of compost. But, at other places, the income from compost is not substantial. Moreover, not all organisations are able to compost bio-degradable waste and hence, at present, income from composting cannot be relied upon as a substantial means of revenue. This must be factored in while making projections regarding the financial sustainability of such projects.

Income groups of service areas: Though not emphasised in the literature available on the subject, this has emerged as the key features of a sustainable community project. On the issue of income group of service areas, the middle-income group habitat has emerged as the common denominator in all these interventions. Of the on-going 15 community interventions, eight are serving in middle and higher income colonies only. The remaining seven are serving all income groups; but they, too, are primarily concentrated in middle income areas.

Higher income areas are generally well served by municipal authorities; in some cases even over-served, as we have found in the case of SMS studies. In the case of low income areas, waste management comes low on the priority as they have to spend on other basic needs first. It is very difficult to motivate them to pay for the services. In middle-income colonies there is demand for these kinds of services but there is lack of requisite supply by the municipality. CBOs/ NGOs therefore find it appropriate to run their operations in middle-income colonies.

While it is not necessary that these interventions are likely to be sustainable only in middle income colonies (many of these interventions are running in all kinds of income colonies), middle income colonies are the common denominator across all the interventions. It appears that it is more feasible to start these interventions in middle income colonies.

The intervention itself should be providing services like doorto-door collection (DTDC) of waste, road sweeping, drainage cleaning, tree pruning, etc. If there are multiple service providers, there should be perfect synchronisation. This improves the public visibility of service areas and keeps residents motivated to continue participating in the intervention. For example, in Mumbai, some ALMS like Diamond Garden have taken upon themselves to arrange for DTDC of waste, drainage cleaning, tree pruning, etc. This has improved the looks of the colony and, hence, residents appeared more convinced about the utility of the interventions.

On the other hand, at CEE's intervention in Bangalore, RWA members responsible for management of the intervention explicitly highlighted the point that due to unsatisfactory performance of municipality in its other civic responsibilities like road sweeping and drainage cleaning, the intervention had failed to make optimum impact.

Any community intervention involves various stake holders. It is necessary to have some sort of formal or semi-formal arrangement which brings together all the stakeholders on a single platform.This helps in the allocation of responsibilities and makes them aware about each other's responsibilities and difficulties involved in their performance. In order to make these interventions sustainable, the study finds that the two most important stakeholders that need to be involved are the local political leadership and the concerned municipality.

Political structure: clearly point out that support of local political leadership is very critical for these interventions; though there is no consensus among community leaders about the involvement of local leadership in these interventions. A community intervention at Vellore had to be abandoned midway because the local council withdrew the land allotted to the intervention. Similarly, at Lucknow, MJS faces a lot of opposition from the local leadership.

The majority of MJS waste collectors are Bangladeshi immigrants and the local leadership considers them a potential threat to national security. At Chennai, it is alleged that the government has deliberately contracted out its municipal services to Onyx in areas where Exnora was working because Exnora was considered close to a particular political party. Whether the active support of the political leadership is needed to run these interventions may be a subject of discussion, but their opposition is definitely troublesome. Hence, in order to make any community intervention sustainable, it is advisable that the local political leadership be kept in good faith.

Municipal linkage: The support of municipalities is crucial to the success of an intervention. The municipality can assist an intervention in many ways: it can issue identity cards to waste collectors which saves them harassment from the police; it can provide land for composting; it can help by performing other services like regular road sweeping, cleaning of drains and local dhalaos (containers).

A municipality's decisions can affect the sustainability of these interventions in very direct ways. For example, at Bangalore, the BMC has started door-to-door collection of waste free of cost. They plan to levy a solid waste tax in the coming months. But, this has started affecting the community intervention in HRBR locality, being run with the help of CEE.

Residents are now reluctant to pay for services that are being provided free by BMC. Had there been an institutional relationship between BMC and CEE, this problem could easily have been avoided. Thus, sustainability of a community intervention requires that there be some sort of forward-backward linkages among the CBO/NGO, municipality and local political leadership. Ideally, all the important stakeholders - CBO/NGO, community, waste collectors, municipality and local political leadership -should be brought together and there should be continuous interaction among them.

Challenges of Community Interventions

Since community interventions in solid waste management involve many stakeholders, any initiative to up-scale these interventions and make them sustainable will have to grapple with numerous challenges, at different levels. Moreover, challenges involved in upscaling these interventions are different from the challenges faced in making them sustainable; though, they are not completely de-linked. Even so, there are some common challenges that can be identified in the up scaling and sustainability of these interventions.

A community intervention cannot be sustainable unless it evokes a wholehearted response from the community. In fact, the real sustainability of these interventions not only involves issues of community participation but of community ownership as well. Non-participating members pose a special problem. They may de-motivate participating households. This, in turn, might trigger a larger demonstration effect, which can affect the sustainability of such interventions. While it is true that some of these interventions are able to sustain themselves on the basis of the user fee collected from the participating households, nonparticipating households could upset the balance.

In other words, in order to be sustainable, a community intervention has to be owned by the community itself. An NGO or CBO, can, at best, be a facilitating agency. Some studies were found that awareness generation among the community is a continuous process. The intervening agency has to be patient and innovative in its approach. In India, the locus standi of Residential Welfare Associations (RWAs) is ill-defined. They do not have legal sanctity and therefore cannot take action against non-participating members. Yet, RWAs are being promoted as a legitimate interface between the community and the outside world by several governments.

In Delhi, all Bhagidari programmes acknowledge RWAs as legitimate stakeholders. There is a growing trend to allocate more roles to these RWAs. The rotation of office bearers also poses many challenges for community interventions. Annual elections of RWAs often result in new office bearers who have to briefed and sensitized. Usually, new elected members have different ideas and solutions for problems. At times, defeated candidates, along with their supporters, create obstacles in the functioning of the newly elected RWAs, affecting the continuity of ongoing interventions.

The different means being adopted by municipalities to implement various provisions of the rules point towards certain up-coming challenges that a community intervention is likely to face in the near future. Municipalities are privatising solid waste services and opting for centralised and mechanized compost plants. For instance, an industrial house has signed contracts with 54 municipalities to help them establish centralized and mechanized compost plants. It seems that the municipalities are not very keen to put in place Decentralised waste management schemes.

The co-operation of NGOs/CBOs is being used to create awareness among the target community regarding waste management. The policy of benign neglect towards decentralised community interventions in itself is a critical challenge for the up-scalability and sustainability of these interventions. For instance, the decision of Bangalore Mahanagar Palika (BMP) to start free DTDC of households waste has severely affected CEE's ongoing intervention in Bangalore. This decision of BMP has made several people withdraw from the CEE interventions which requires them to pay a user fee.

The unsatisfactory, or below-par, performance of municipal staff can create challenges even for institutionally linked community interventions. This is evident from the experience of Pramukh, Dehradun and NBJK, Ranchi. Pramukh was permitted to dump the waste in municipal bins from where the municipal staff would take it to the landfill. A strike by the Dehradun Nagar Nigam staff posed a new kind of challenge for Pramukh as the waste piled up in the bins. At Ranchi, NBJK workers used to supervise the works of Ranchi Nagar Nigam's (RNN) sweepers and drainage cleaners.

To protest against this external supervision, the RNN staff went on strike. As a result, RNN asked NBJK to stop on-site supervision of its staff. Besides these challenges, community interventions have to deal with many other issues that affect their smooth functioning and have an impact on their sustainability as well. At some places, community members expressed their concerns over the frequent turn over rate of waste collectors. According to them, it disrupts the continuity and the waste collectors have to be instructed and trained afresh as per the residents' preferences. Some women residents said that this frequent turnover of waste collectors creates a sense of insecurity too in affluent urban residential areas.

Challenges in Upscalability

Some of the basic challenges involved in upscaling/ replicability of community interventions are:

— Every new intervention may have to start afresh. Geographical locations, community composition and its income category may all be different, though, the previous experience will definitely provide a road map to proceed farther and faster in terms of output.

— Change in the physical location - even a change in the municipal area within the same municipality - might create challenges. For example, the new municipal authorities may not be as receptive to such interventions.

— Different habitats require different approaches. For instance, private residential colonies, government staff quarters, educational institutions, working offices, etc, all have different rules and regulations governing the conduct of their inhabitants.

This, in turn, engenders different kinds of awareness and behaviour patterns among people that necessitates different kinds of interventions at different places.

Waste Collectors

Waste collectors are a key component of a decentralised solid waste management system. Working in the informal sector, they treat about 10 to 15 % of the total waste of a city. There are serious occupational and safety hazards associated with the work. In addition they are subjected to different kinds of social and economic harassments, especially in the case of women waste collectors. Here, is a situational analysis of waste collectors working in the community projects.

Earning More Money

Earning more money is definitely a basic motivation, but it is not the most important motivation, especially in the case of socially low ranking occupations, as is depicted in the diagram. Here, 49% waste collectors claimed that they were earning less money than in their previous work situation, or when they were working as informal waste collectors. Some of the important reasons cited by waste collectors to continue working in community interventions despite earning less are:

— *A distinct identity*: The practice of providing uniforms to waste collectors has helped them acquire respectability in the eyes of the community.

— *Assured livelihood*: A monthly salary guarantees a constant flow of income and protects the waste collectors from a fluctuating income which is inherent in informal work; though, at times, they may be earning less.

— *Reduced harassment*: The formal or informal arrangement of community interventions with concerned authorities has provided legitimacy to the work of waste collectors. As a result the harassment that they had to face at the hands of municipal and police personnel has ended.

— *Greater social acceptance*: Waste collectors were of the opinion that the community understands their contribution better and values their work.

Muncipality

Despite all the efforts of community organisations to formalise the work of waste collectors there is still a continuing degree of informality associated with their work.

This is evident from the method of recruitment of waste collectors. The methods of recruitment of waste collectors are quite informal. 60% of waste collectors got their present job either through their relatives and friends or by their own efforts. Only 36% were employed due to efforts of the service providing organisations. The remaining 4% were recruited on the recommendations of RWAs and municipal staff.

Moreover, majority of them (74%) are working as waste collectors for the first time. Hence it is not that all ragpicker livelihoods are substituted with that of a formal waste collector as there has been no specific attempt to recruit them on a priority basis. This is understandable since the waste picking is a low skilled job and can be learned easily. Though community organisations claim to train them before appointing, survey reveals that only a small percentage of the total respondents have been given any kind of training in waste management.

The MSW Rules, 2000, were to be implemented in letter and spirit from January 1, 2004. This timelines was expected to create a sense of urgency among municipalities to implement the various provisions of the rules in a timely fashion. Surveys reveals that even after a three-year grace period solid waste management has a low priority. This is evident from the following:

— There has been no departmental restructuring to streamline solid waste management functions. Most municipalities reported that there had been no internal capacity building to develop requisite skills to implement these rules.

— Even the responsibilities for solid waste management vary from place to place. At some places it is the Health Commissioner who holds the ultimate responsibility for these functions, while at other places it is the Chief Civil Engineer who is responsible for SWM. It appears, from general discussion with municipal officials and lower level staff, that there is no clearcut division of roles and responsibilities.

— It seems that there is no provision to have separate budgetary allocations to perform solid waste functions. Moreover, even if there are specific funds allocated for this, they remain unspent. For example, municipalities in Jharkhand could not spend a fund of rupees 8 crores specifically allocated for SWM services in the state.

— There is a prevalent belief among municipalities that the deadline for implementation of these rules will be extended. Moreover, some municipal officials of small cities point towards metropolitan cities which have better resources and have still not been able to implement the rules within the given time frame.

House-to-house Collection of Waste

Out of 31 municipalities surveyed so far, 14 have initiated household collections of waste, though not at the entire city level. Again, it needs to be mentioned that these municipalities are at different stages of primary collection of waste. Suryapet, Nasik and Panjim have reported more than 90% door-to-door collection.

Bhadreshwar and Kalyani municipal authorities have initiated household collection of waste in the entire municipal area. On the other hand, municipalities like Mumbai, Delhi, Chennai, Bangalore, Nanded and Indore have started household collection of waste in some parts of their service areas. Rest are yet to start primary collection of municipal waste.

Privatisation of Municipal Services

Privatisation of solid waste services by municipalities needs a careful examination. At stake are several issues such as livelihoods of waste collectors, 3R principles of waste management (Reduce-Reuse-Rrecycle), cost-efficiency of services. All of these have a critical bearing on sustainability of these services as well as of the environment. In order to critically examine the privatisation issue, it is important to review some of the privatization efforts.

In Nanded, the corporation has signed an MOU with CDC for collection, segregation and transportation of some portion of the city 's solid waste. As per the agreement, the municipality will pay a sum of Rupees 476 per house per annum to CDC for collection and transportation of solid waste from households. The corporation plans to levy a solid waste tax to recover this cost. This contract will be renewed every year after a performance review of the services. For this, a complaint book is maintained, which needs to be filled up by residents every month.

In Nasik the collection and transportation of the city waste has been leased out to a single private operator named Adarsh Ghanta Gadi Prakalp. Nasik Municipal Corporation pays Rupees 475 for every tone of garbage collected and transported up to the landfill site. The contract is to be renewed every year subject to the satisfactory performance of the operator.

The Surat Municipal Corporation has contracted door-to-door collection to three private operators, namely: CDC of Jaipur, Jigar Transport of Surat and Global Waste Management Cell (GWMC) of Mumbai. Each contractor is to be paid at the rate of Rupees 645 per tonne of garbage collected from households and transported to the transfer station. The contract has been signed for seven years but there will be a continu-ous monitoring of the performance of the private operators by municipal officials. A different contract has been signed for the transportation of waste from the transfer station to the landfill at Khajod. The contractor is paid Rupees 7.71/km/ tonne of waste transported.

New Delhi: A consortium of three companies, Subhash Projects Marketing Limited, Dooars Transport and Tetratech India Limited would be responsible for collection, segregation, storage and transportation of Delhi's central and south zones respectively. At present, MCD will pay Rupees 600 per metric tonne of garbage transported to the landfill site. The payment will be made after a third party verifies the expenditure.

The privatisation of waste collection, segregation and transportation raises several critical questions:

— The practice of making payments on the basis of the amount of waste collected creates a bias in favour of more waste creation. The contractor makes more money if there is more waste. This practice ignores the basic principles of waste management - reduce, reuse and recycle. Even in the cases where payment is made on the basis of number of households covered; there is no incentive for the contractor to make efforts to reduce the waste generation. Most importantly it negates segregation which in turn negates better recycling and composting.

— Where the contract is renewed every year, it is detrimental to the interests of both parties. For instance, in Nasik, the entire waste collection of the city requires more than 110 vehicles and a corresponding workforce; this is a huge infrastructural investment which is very difficult to arrange by another contractor in a short time span. This might lead to complacence on the part of the private operator.

— The system of monitoring also leaves much to be desired. Monitoring is to be done either by municipal staff or by a third party, as in the case of MCD. This monitoring system has failed to provide any roles to community or community-based organisations. This inverts the logic of implementation of MSW Rules. The implementation of these rules demand that residents should do source segregation and NGOs/CBOs should be involved to create awareness among residents.

— Above anything else, this privatisation process has the potential to work against the interests of the traditional waste collectors. The costbenefit calculations of these contracts motivate private operators to adopt more echanised and high-tech equipments, which necessarily entails reduction in labour costs. This might render several waste collectors jobless.

Composting System

In India, the need for composting biodegradable waste is much more pronounced since it constitutes about 60% of the total municipal waste. There are two types of composting systems prevalent in the country. On the one hand, municipalities have established capital-intensive technology-based centralised compost plants where the mixed waste is composted. For instance, Municipal Corporations of Delhi, Nasik, Jalandhar, Trivandrum, Calicut, Bhopal and Ahmedabad have established centralised and mechanised compost plants under different arrangements. On the other hand, several communities are engaged in locally composting their bio-degradable waste by adopting indigenous and labour intensive technologies. Here, we have examples of Exnora Green Cross Vellore, Jan Chaithanya Vishakhapattanam and Stree Mukti Sanghathan Mumbai.

Centralised Composting Systems

Information about mechanised compost plants was difficult to obtain due to the reluctance of municipal and plant authorities to share any information with the research team. However, we do have relevant data of the compost plants in Ahmedabad, Nasik, Bhopal, Gwalior and Jalandhar. All the plants produce compost from mixed waste.

In Ahmedabad, Excel Industries has set up a compost plant with an investment of about Rupees 6-7 crores. The plant has been able to reach a breakeven point in four years. In Nasik, the Municipal Corporation has set up a compost plant at the cost of Rupees 5 crore. Though the plant is running, its long-term sustainability is questionable. The compost is produced at a cost of Rupees 2,500/tonne whereas it is sold at Rupees 1,700/tonne. Owing to this fact, Leaf Biotech Private Limited, a Thane based firm, which had established and run the plant for two and a half years, has backed out of the contract.

In 1993, M. P. State Agro Industries (MPSAI)had set up two compost plants at Bhopal and Gwalior. As has been the pattern across the country, these two compost plants were set up with the technical help of Excel Industries. The compost produced is being sold under the brand name of 'Agrorich'. But, MPSAI had to shut down its Gwalior compost plant since it could not find a market for its compost. Till last year, the Bhopal plant was also running at a loss. This year, it has been reported that the plant will make some profit as it has received a bulk order for 5,000 tonnes of compost from a Jaipur-based firm. According to the agreement the plant will have to supply the compost under the brand name of 'Croprich Gold'.

The agreement was originally signed between the Jalandhar Municipality and Excel Industries. Later, Excel Industries sub-contracted the compost plant to Punjab Grow More Fertilizers Limited, Jalandhar. The brand name of the produced compost is Shakti Jaivik Khad. This plant is also running at a loss. To sum up, except the Excel plant at Ahmedabad, no other mechanised compost plant has reported breaking even. As the Nasik plant's figures show, the production cost is greater than the selling cost. With establishments costs of these centralised compost pants running into crores and their economic sustainability yet to be demonstrated, there is a need to re-consider the decision of municipalities opting for centralised compost plants.

Decentralised Compost Systems

According to data collected during the documentation process, 11 out of 16 community based interventions are composting bio-degradable waste. Out of these, Exnora Green Cross Vellore (EGC) and Stree Mukti Sanghathana(SMS), Jan Chaitnya stand out for their innovative approach and the amount of bio-degradable waste composted. SMS workers compost about 20 tonnes of bio-degradable waste daily.

SMS, with the help of ALMs, has been able to construct and run compost systems on roadside drainages and even in multi-storey apartments. SMS, has trained its waste collectors in composting and in the process has provided them with an extra source of income. EGC has set up compost sheds in places like fish markets, temples and on land allotted by the municipality or panchayats. It uses cattle dung to expedite the process of decomposition. EGC, at its Palavansethu village intervention, has produced compost worth more than Rupees 800,000 and has tied up with the district forest department to sell compost manure.

Other community interventions have not been so successful in marketing their compost. MJS was forced to discontinue its compost because its tieup with the Central Government fell through and it could not find an alternate market. However, since the establishment costs are low for these decentralised compost systems, they do not run the risk of running into losses.

Composting Challenges

— *Infrastructural problems*: The most important requirement for a composting site is land which is high-priced in an urban area. Government agencies that own most land in urban areas usually do not work in coordination. Siting a compost plant at a socially acceptable place is also a challenge. The NIMBY (Not In My Back Yard) syndrome has to be dealt with tactfully.

— *Procedural and technical problems*: All the processes and technologies used in composting cannot succeed in producing good quality compost unless they get segregated waste. The myth of odour in composting is more of maintenance problem than a process problem.

— *Quality assurance*: The quality of compost is generally based on the nature of bio-degradable waste. The presence of weeds, heavy metals and other objects, and the technologies used in composting, all have a bearing on its quality. Compost plants also reported that their operation and maintenance costs escalate due to mixed feed stock. In the absence of primary source segregation in India, producing good quality compost is challenging. Without government guidelines and a certification authority there is no way of judging the quality of compost.

— *Marketing challenges*: To market compost at affordable as well as economically viable price is another challenge, which is inextricably linked with issues of quality control and government policies. In absence of any certification authorities, compost makers are not able to put up their case strongly before the consumers, who have more faith in scientifically certified chemical fertilizers.

Moreover, these fertilizer units are provided with government subsidies, which enable them to remain in the market while compost, which is organic by nature, fails to find market for itself. In order to create a level playing field, it is necessary that compost products should also be provided subsidy and government should set up a new agency to specifically cater to and promote compost manure.

Model Development

Considering India's diversity - cultural, socio-political,

geographical and economical - it is unlikely that a single model of a community-based solid waste management system will be applicable to all areas. Still, such an exercise can provide a benchmark against which we can measure field interventions. Any such exercise will necessarily have to be based on some assumptions, which, in order to be realistic, will need to take into account the ground realities.

Having collected data from 25 community and municipal interventions in solid waste management across the country, the research team has attempted to make such assumptions. Across the world, it has been found that when solid waste management is being provided at the community level, the economies of scale are not so pronounced. But, as per the Full Cost Accounting Handbook of EPA, no economics of scale are thought to exist for communities comprising more than 50,000 people. Given the joint family tradition in India, if we assume that there are five members in each family, it will equal to 10,000 households.

Moreover, this assumption seems valid against the actual field data. For instance, as per the calculations of Kancharapara municipality, a solid waste management intervention covering 14,000 households will need an initial investment of about Rupees 122 lakh. Obviously, this scale of investment cannot be expected from community level interventions. Hence, we can assume that community based interventions may be considered up to a maximum of 10,000 households: though, even this will require a substantial investment.

Cost Calculation of Solid Waste Services

Ideally, there are two ways through which one can calculate the cost of solid waste services in a community: activity-based calculations and pathbased calculations. MSW activities are:

- Collection of waste
- Transportation of waste to transfer station
- Processing/disposal of waste
- Any sales of recyclables or compost

Solid waste management paths are:

- Recycling
- Composting
- Waste-to-energy
- Land disposal

However, in actual practice, these categories are not discrete; there is an overlap among them. In order to keep calculations simple, it has been decided to break these activities and paths into two categories: costs likely to be incurred and revenues likely to be generated. In an ideal decentralised community-based solid waste intervention, the elements of costs will be as follows:

- Collection cost
- Transportation
- Operation at transfer/disposal station: this includes operation and maintenance of compost sheds too.
- Awareness materials/ trainings of personnel.
- Organisational expenditure.

Elements of revenue will be as follows:

- User fee
- Sales of recyclables
- Sales of compost

In order to arrive at cost-benefit calculations for a 10,000 household solid waste intervention, we need to make following assumptions. As already mentioned, most of these assumptions are based on actual ongoing filed experiences that have been documented. Some of the basic assumptions are as follows:

- All likely costs and revenues are calculated at the present and fixed value.
- Land and shed for segregation and composting has been provided free of cost by municipality/ some other agency.
- Kinds of ser vices offered: door-to-door collection and sweeping of nearby roads.
- One waste collector collects waste from 150 households, sweeps the nearby roads and then works on the compost shed.
- There is one supervisor for every six waste collectors.
- Two waste collectors share one rickshaw trolley covering 300 households.
- Each waste collector is paid a salary of rupees 2,400 per month. Basically, his/her salary is rupees 1,800 per month. The remaining 600 are being given for recurring expenditures like primary medical health, maintenance of rickshaw trolleys and uniforms and other equipments given for the waste collection.
- Each supervisor is paid a monthly salary of rupees 3,000.
- Only 80% of the households are expected to pay the user fee.
- Each Ricksaw trolley costs about Rupees 9,000.
- Each waste collector is provided with one time inventory worth rupees 2,500. This includes items like uniform, gloves, shoes and other equipment.

Based on the above assumptions, we can calculate the likely expenditure to be incurred and revenues likely to be generated for any number of households. Let us first, attempt this calculation for 10,000 households.

Cost Estimation

For reasons of simplicity, it will be better to first calculate one-time establishment charge and then calculate the recurring monthly expenditure. Onetime establishment cost can be broken down into following cost elements:

1. Expenditure on Rickshaw Trolleys: Number of rickshaw trolleys needed is 10,000/300= 34. Cost of one trolley=Rupees 9,000. Total expenditure on rickshaw trolleys= 9,000 x 34 = Rupees 3,06,000.
2. Expenditure on accessories for waste collectors: C o s t on accessories such as uniform, gloves, boots, etc worth Rupees 2,500. Total expenditure=2,500x67=Rupees

   1,67,500.
3. IEC material/ awareness workshops for the community: Assuming that at if at least 40 awareness workshops, catering to 500 residents per campaign, are organised and expenditure incurred on each workshop is Rupees 5,000, the total expenditure on awareness workshops of residents will be Rupees 200,000. Finally, we can assume that there will be an expenditure of about Rupees 1,00,000 in publication and distribution of the IEC material. Thus, the total expenditure incurred under this head will be (2,00,000+1,00,0000)=Rupees 3,00,000.
4. Training of waste collectors/ supervisors: Assuming that all the waste collectors and supervisors are trained for a minimum of 30 days and they are paid as per their monthly salary, total incurred expenditure will be Rupees 1,96,800.
5. Organisational overheads: One-time establishment cost of intervening organisation can be safely assumed to be

about Rupees 40,000. Total one time establishment cost (in Rs.) is 1+2+3+4+5=10,10,300.

Recurring cost per month: Having calculated the one-time establishment cost, now we need to calculate the recurring expenditure per month, since salaries to the workers will be paid on a monthly basis and the user fee will also be collected on a monthly basis. The elements of recurring cost per month will be as follows:

1. Salary to waste collectors = 67 x 2,400 = Rupees 160,800.
2. Salary to supervisors = 12 x 3,000 = Rupees 36,000.
3. Organisational expenditure = Rupees15, 000.
4. Incidental expenditure = Rupees10,000.

Based on the above assumptions, the cost-revenue calculations, for different number of households can be put down into tabular forms. Now, all these elements of costs and revenues can be put up into a simple linear equation to calculate the time period needed for the attainment of breakeven point of the intervention and the subsequent profits thereof. If we denote:

— Recurring monthly expenditure as Mc
— Revenue from sales of recyclables as Rr
— Revenue from sales of compost as Rc
— evenue from user fee as Rf
— Total establishment cost as Ec.

If the intervention attains the break-even point in n months, then this equation will be:

$$Rf \times n + Rc \times (n-3) + Rr \times n = E_c(1+r^*/12)^n - Mc \times n$$

*r is the rate of interest per annum, which has been assumed as 6 percent. Based on the above equation, breakeven periods (in months) for different number of households can be presented in the following tabular format:

Number of households	Break-even period (in months)	
	No Revenue from compost	Recyclables to workers and user fee@35
10,000	17	20
5000	24	31
2000	44	63

Benefits of Decentralised Solid Waste Management Systems

The different models discussed above have different spin-offs, though many of these are common across the models. However, here, we will be primarily concerned with the ideal model identified by us which would be applicable in practice as well. The ideal model will be one in which:

— There is source segregation of waste.

— There is door-to-door collection of waste.

— There is no mechanical vehicle involved in primary transportation of waste. They are used only for secondary transportation (to transfer the inerts and other remains from the composting and recycling shed up to the landfill site by the municipality).

— Biodegradable waste is being composted.

Apart from providing a sustainable solution to waste management, this system has many direct as well as indirect economic, social, health and environmental benefits. Some of the important benefits are:

Economic benefits

— Livelihood creation: As has been described earlier, this model is labour intensive rather than capital intensive. Thus, a solid waste management system of this kind covering a household population of 10,000 has the potential to provide employment to 67 waste collectors, 12 supervisors and 2-3 persons of the intervening organisation.

— Source segregation keeps the recyclable material cleaner, which, in turn, fetches higher prices.

— Consequently, the quality of end products made from these recyclables improves many folds, which in turn, fetches higher prices and helps in preserving and promoting the faith of committed consumers in these recyclable goods. Plus, it can give a new fillip to the recycling industry.

Economic benefits for municipality

— Municipality can save up to Rupees 4,32,000 per annum in secondary transportation for a 10,000 household programmes.

— Municipality can save labour cost, as there will be no need for it to employ people for primary transportation of waste.

— It will reduce the burden of municipal staff as there will be less need of regular road sweeping and drainage cleaning.

High landfill diversion rate: Apart from offering direct economic benefits, the model offers many types of indirect economic benefits. For instance, this model succeeds in attaining a landfill diversion rate of more than 80%, which not only saves money in terms of excess land to be acquired for the new landfill site but also conserves such a natural resource for some other useful work.

Health benefits: The provision of formalising the working conditions of waste collectors provides them with the opportunity to work in healthier conditions. The provision of gloves, uniforms and other safety equipment improves their working condition. The neat and clean neighborhood makes the area less prone to diseases. The reduction in number of mechanised vehicles used for primary transportation of waste results in reduced emission of many harmful gas, which indirectly benefits the health of the all the residents of the city.

Social benefits for waste collectors: The waste pickers could be substituted as waste collectors and their livelihood would be formalised. They get better recognition and dignity by working as formal waste collectors than as waste pickers. There is reduced or no harassment by municipal staff and police working as formal waste collectors.

Compost: The practice of making compost not only provides an extra source of revenue for the system but also helps reclaim the lost fertility of the soil. The use and abuse of chemical fertilisers are well known and promotion of compost as a natural manure is a pressing need of the time.

Decentralised solid waste management systems, premised upon the management and ownership of local people, have a lot to contribute to the strengthening of civil society and will result in creation of a much more aware and empowered citizens, who will carry forward these new skills in various other walks of life.

5

Optimising Wastewater Treatment

The municipal sewage system and industrial wastewater treatment are inadequate in many urban situations. A municipal sewage network may be in place, but coverage is usually incomplete, and the level of treatment provided is inadequate. Even where reasonable treatment facilities exist, poor maintenance and operation often result in failure to meet design effluent levels.

In such circumstances, management of industrial wastewater discharges is also frequently poor, with uncontrolled discharges of untreated effluent to surface waters (through often drainage or stormwater channels) or to the sewer system. The result is high levels of water pollution. It is not uncommon for streams or water bodies to be almost or completely anaerobic and heavily polluted with organic compounds, pathogens, and heavy metals. There are several objectives that must be addressed in such a situation:

— The collection and removal of domestic and municipal wastewater to protect public health and to improve the immediate environment (particularly important where inadequate disposal is resulting in groundwater pollution)

— The establishment of an effective industrial pollution control system to reduce the loads and impacts of industrial discharges

— Provision of municipal and industrial treatment as necessary to protect the environment at the points of final discharge

— Efficient and cost-effective achievement of all these goals within the relevant social and political constraints.

From the environmental (as distinguished from the sanitation) point of view, the focus must be on the receiving water bodies. The problems are typically diffuse, with hundreds or thousands of small discharges and with the problems concentrated to some extent where particularly polluted streams or poorly treated effluents discharge to major water bodies. Upgrading or extension of the wastewater collection system may reduce this diffuse pollution but may produce major point discharges that must receive adequate treatment. A wastewater strategy must therefore be based on a water quality plan for all the receiving waters in the catchment, usually on the basis of water quality objectives.

It is necessary to have explicit medium- to longterm objectives for the quality of water in the various water bodies in the catchment under consideration. These objectives are often based on defined beneficial uses for the water bodies, typically including about a half dozen uses such as source of water supply, agricultural use, fisheries protection, and so on. A set of key numerical parameters can be defined for each use, and the water quality objectives can be developed in terms of uses for different sections of the water bodies and a strategy for achieving those standards. The objectives then provide clearly defined goals for protection or improvement of each section of the system.

The basic responsibility for municipal sewage lies with the government. Industrial wastewater treatment is fundamentally the responsibility of the enterprise but in practice has to be driven by government action. The challenge for the government is to use the whole range of options and instruments available to achieve the objectives outlined above, combining physical and operational improvements in the municipal infrastructure with the controls and incentives necessary to induce improvements in the industrial sector.

Strategy Development

Load Estimation

The first step in developing a wastewater strategy is to estimate the overall loads in the catchment over the time scale being considered, which is typically about 20 years. This will require, in addition to information on population growth and densities, estimates of industrial activity and of projected changes in industrial and population patterns. In some cases, direct observations of industrial pollution loads are available, but more often, estimates are based on statistical information on economic activity (sectors, employment, turnover, and so on), using various coefficients for the unit loads of pollution.

Overall planning requires estimates of both domestic and industrial loads on a geographic basis and over the time period under consideration. The estimates need to be developed for key parameters such as suspended solids, oxygen demand, nutrients, organic materials, and heavy metals, depending on the particular characteristics of the catchment and receiving waters.

Reductions Determination

Once load estimates are available, it is possible to determine the reductions in present and future loads needed to achieve the water quality objectives. In simple cases, a mass balance may suffice, but often it will be necessary to carry out water quality modeling. The objective of the modeling is to estimate the impacts of the increasing loads on water quality and to identify where load reductions are required in order to achieve the water quality objectives. The sophistication required in the modeling will depend on conditions. In some cases, a simple one-dimensional model of oxygen depletion will be acceptable; in other cases, complex models will have to be developed to address water circulation and the degradation and interaction of several pollutants.

Options for Load Reduction Development

After the desired degree of reduction in pollutant loads has been estimated, the next step is to develop options for achieving that reduction. If the most significant pollutants are those associated with industrial effluents-for example, complex organic compounds or heavy metals-the control efforts will clearly be concentrated on the industrial discharges. Often, however, oxygen depletion and nutrients are the critical issues, and the causes are typically a mixture of municipal and industrial sources. Then it is necessary to control both types of sources. The costs of cleaning up a major industrialized urban area can be massive.

The estimated costs of water pollution control in Shanghai in 1986 were US$1.4 billion. Preliminary estimates show that the Buenos Aires sewerage authority faces an investment program of nearly US$1 billion over the next decade. Clearly, such programs require decades for implementation, making it important to tackle them in an organized and cost-effective manner.

Components of Urban Wastewater Program

An urban wastewater program comprises several distinct but interlocking components. Municipal system improvement is almost always a central feature, but the emphasis given to the industrial wastewater control component depends greatly on the extent of the industrial contribution to the overall problem, the types and sizes of industries involved, and the costs of enforcement and implementation. In some cases, or for some pollutants, small or nonpoint sources may be a significant problem, and one that is typically difficult to tackle.

Upgrading of Municipal System

There are normally two imperatives behind municipal system upgrading:

— Expansion of the coverage and quality of sewerage provision

— Reduction of the impacts of final disposal of treatment plant effluents.

However, because of limited funds, sewerage authorities often have to make tradeoffs between expanded coverage and higher levels of treatment, with consequent implications for the quality of the receiving water. The impacts of final disposal depend, obviously, on the discharge location. In many cases, an existing system configuration more or less limits the choice of the discharge site, and therefore the emphasis is on improving the level of treatment provided.

Municipal wastewater systems are normally designed to treat influents that are essentially domestic in nature. Such systems are ineffective in removing some industrial pollutants and may even be damaged by them. Design of municipal wastewater treatment is a sophisticated operation. In general terms, however, there are three major types of process, in ascending order of removal effectiveness (and cost): physical, sometimes assisted by chemicals; biological; and "advanced," which includes further chemical or biological stages, filtration, or combinations of these methods. These systems can achieve high levels of removal of organic material and of suspended solids. The advanced systems can also remove nutrients to a high degree. Municipal systems do not cope well with high concentrations of complex organic chemicals such as solvents and hydrocarbons or of heavy metals.

The removal efficiencies are low, and biological treatment systems can be poisoned if incoming levels are too high. Other wastewater treatment processes that can be tailored to deal with such industrial effluents are available. Because of the limitations of municipal systems, and to protect the physical infrastructure and workers, it is normal practice to require pretreatment of industrial effluents that are discharged to a sewer system.

Control of Industrial Effluents

Treatment systems for industrial effluents can be designed to provide any required level of pollutant removal, although at increasing cost and sometimes with a resultant wastewater treatment sludge that presents its own disposal problems. Where effluent treatment costs are high, waste minimization programs become very worthwhile. The degree of industrial effluent treatment required is established, in theory at least, in relation to relevant ambient quality or effluent standards. In practice, control of industrial effluents is frequently poor, and industry may be a major contributor to the overall pollution load. Where practical controls exist, industry is typically faced with two choices: direct discharge to surface waters (licensed groundwater discharge is rare), or discharge to the sewer system, if one is available.

Effluent standards will apply to both options. Sewer regulations will require pretreatment to remove toxic substances, but effluents that can be treated by normal municipal systems will be accepted, at a charge. Direct discharge standards will depend on the character and objectives of the receiving water but would normally be expected to be more stringent than sewer standards. Because of economies of scale, sewer discharge of simple wastes such as BOD is often cheaper than industrial onsite treatment. However, there are often problems with the capacity of the municipal treatment system and with implementing correct charging procedures, and so this option may not always be available. Clearly, where regulations are inadequate or enforcement is lax, there is a financial incentive for industry to avoid treating the effluents.

Optimising the Program

Once the basic information on water quality, municipal and industrial loads and trends, and estimated control costs is available, it is possible to begin to optimise a wastewater management program. A key decision variable is the time scale over which the required upgrading is to be implemented. The costs of major treatment systems are so high that upgrading almost always has to be staged.

Moreover, high urban growth rates mean that significant investment is often required just to maintain present levels of service to the growing population. Implementation of effective industrial pollution control programs takes time, and a realistic approach to projecting load reductions must be adopted. An iterative planning process is therefore required that examines a number of options for the scale and rate of wastewater treatment improvement, balancing the costs of the program against the time needed to achieve the water quality objectives. This process should involve an appropriate level of public discussion so that a practical program can be developed that will have the broad public and political support necessary for implementation.

Costs and Benefits

A set of agreed water quality objectives (WQOs) that has been adopted by the government can be taken as reflecting the value of improving the receiving water quality, assuming that it is based on evaluation of the economic benefits of the improved uses of the water resources and on the outcome of a public priority-setting process. The major components of a wastewater management plan, which typically compete for investment funds, are:

— Upgrading of sewer systems in existing urban areas to remove pollution from neighborhoods and to reduce uncontrolled discharges to local watercourses and groundwater

— Upgrading of municipal treatment systems to reduce the impacts of the effluent discharges on the receiving waters

— Introduction of a system to identify and regulate discharges from industry

— Reduction of current industrial pollution loads through recycling, improved waste management, onsite treatment, or connection to sewer systems

— Adequate provision of sewerage and treatment for new urban development

— Effective control of effluent discharges from projected new industrial development

— Development of programs to quantify and tackle nonpoint sources of pollution, including combined sewer overflows.

Both the overall costs of these components and the distribution of costs must be taken into account in arriving at an estimate of the most cost-effective investments for achieving WQOs. In effect, a marginal cost curve can be developed for the water quality improvements, although there are always many uncertainties in the estimates. Two practical problems have to be resolved in preparing realistic options: the actual costs of pollutant removal for each component and the impact of such removal on water quality.

Unit Costs of Pollutant Reduction

Each of the components outlined above will have a different effective cost of pollutant reduction, and the distribution of the burden of the costs will usually be different. For example, for BOD, which is usually one of the main parameters, the following general conclusions can be drawn. Upgrading sewer systems can greatly reduce local pollution loads but will increase the loads at the treatment plant. It is reasonable and realistic to set domestic charge levels to cover at least this component of sewerage, since it provides direct benefits to households. Thus, it should be possible to cover investment costs out of increased revenue.

Upgrading municipal treatment addresses what is often the single largest point source of BOD in a system, and the costs of removal can be calculated quite accurately. Sludge handling and dis-posal costs can be a significant element and must be included in the estimates. BOD removal normally entails increasing marginal costs in moving from primary to secondary systems and on to advanced systems.

The costs of treatment should, in principle, be borne by the system users (the polluters pay). It is often politically difficult, however, to raise surcharges enough to cover the higher treatment levels because the users do not see the benefits directly. In many projects, some component of the treatment costs is borne directly by the government. Introducing industrial pollution controls is used to achieve reductions in industrial effluent discharges. To do this, a regulatory and permitting system has to be in place, whether it is based on standards or on charges.

The cost of putting a system in place or reinforcing it is part of the investments that are necessary (but not sufficient) to achieve reductions in industrial discharges of BOD or other pollutants. The design of the system should specifically address how effective it can be in actually achieving certain levels of reductions. This effectiveness depends on a number of factors, but the number and size of polluters is clearly a key one: it is much quicker and more cost-effective to deal with a small number of large firms than with many different small ones.

Reducing industrial loads can often be done at little or no net cost to industry, even for significant reductions, but there are often transaction costs that are typically borne by the government. Estimates can be made of, for example, the volume of BOD generated by industrial sources and the costs of reduction, if an inventory of sources is available. Clear priority should be given to ways of inducing waste minimization as a first step in reducing overall loads.

In principle, the costs of treating BOD loads from industrial sources should be no more than the costs of municipal treatment because industry can, in the ideal case, choose to use the municipal sewers and pay the costs. Given the waste minimization opportunities that typically exist in industry, the marginal costs of pollution reduction should be no higher than the costs in the municipal system. New urban developments should be provided with sewerage and treatment systems adequate for meeting the necessary discharge requirements. The costs should be borne by the users.

In practice, however, fringe developments are often expensive to service and are occupied by poorer (often illegal) households. Projections of development should therefore include realistic estimates of the extent and net cost of control of expanding urban areas. New industrial development presents a much easier task in enforcing effluent standards than does retrofitting older plants. The net cost of controlling new pollution loads can therefore be expected to be less. It is important, in setting water quality objectives, to take into account the growth of urban and industrial activity so that realistic discharge requirements can be placed on new projects. Nonpoint sources account for a significant load of many pollutants, including BOD but particularly nutrients. This category typically includes runoff from urban and agricultural land but can be broadened to include small polluted urban drains and streams, where the precise sources of the pollution are too small and numerous to be readily identified.

The costs of controlling these sources are typically high. Unfortunately, the loads may be also high, so that it is difficult to achieve water quality objectives by dealing with point sources only. It is therefore important to try to address the extent and control the costs of nonpoint sources. From detailed analysis of the sources and the costs, it is possible to estimate marginal reduction costs for the major types and locations of pollutant loads. These load reductions must then be translated into real water quality improvements.

Optimising Load Reduction

Most large water catchments are not uniform and fully mixed, and therefore not all load reductions will have the same impact on final water quality. In most cases, too, the WQOs vary across the catchment. It is therefore necessary to estimate (usually using a water quality model) the improvements that can be obtained with different levels and locations of load reduction. For some pollutants, such as heavy metals, the number and location of sources may be sufficiently limited that such modeling is not required.

The model makes it possible to identify, to an acceptable level of uncertainty, the most cost-ef-fective investments for achieving the desired WQOs. Once an initial estimate has been prepared, one can examine the implications of adopting more or less ambitious objectives. On this basis it is possible to carry out an informed process of discussion and agree on a water quality plan and a wastewater management strategy and program. The approach outlined here is standard when the problem is presented and tackled as a water quality management issue.

Unfortunately, in sector projects, such as municipal services, industrial upgrading, and pollution reduction projects, the tradeoffs between the different water pollution sources are sometimes not recognized. Priority investments were identified by a screening process, taking into account the size of the load, the cost-effectiveness of the actions, and the impacts of different types of pollution.

Two perspectives were used to evaluate cost effectiveness: regional benefits at the level of the Baltic Sea, and benefits to the local population and environment directly affected. Most of the actions identified were cost-effective at both levels, but the priority ranking on cost-effectiveness differed. For example, the cost of reducing loads on the Baltic Sea varied from 8 European currency units per kilogram (ECU/kg) for the most cost-effective plant to 21 ECU/kg for the project ranked ninth. The recommended priority investments were based on a balance of local and regional rankings.

6

Biodegradation

Over time, a compound will biodegrade at a particular rate and the biodegradation kinetics will be dependent on the environmental conditions and the availability and concentration of the substrate. The Monod equation was developed to describe the growth of a population of microbes in the presence of a carbon source. At low concentrations of substrate, the microbial population is small. With increasing substrate concentrations, the microbial population grows until a maximum growth rate is reached.

Both first and zero-order rate constants are calculated when little to no increase in microbial cell numbers is seen. This will occur where the cell density is high compared to the substrate concentration. In this case, biodegradation kinetics are better represented by the classic Michaelis-Menton equation for enzyme kinetics.

This equation assumes that the reaction rate of the individual cells and not the microbial population is increasing in relation to increasing substrate concentrations: where v = reaction rate (μ in the Monod equation), V_{max}=maximum reaction rate (μ_{max} in the Monod equation), and K_m is the Michaelis constant (K_s in the Monod equation).

Zero-Order Rate Constants

A zero-order rate constant is calculated when the substrate concentration is much greater than K_m so that as the substrate is biodegraded, the rate of biodegradation is not affected, i.e. loss is independent of substrate concentration. The rate of a zero-order reaction is linear (a constant amount of the substrate is lost per unit of time) and is represented by the differential:

$$\frac{dS}{dt} = k_0 \tag{1}$$

and the integral:

$$k_0 = \frac{S_0 - S}{t} \tag{2}$$

where S_0=initial substrate concentration, S=substrate concentration at time=t, and k_0=the zero-order rate constant (expressed as concentration/time, e.g. µg/L/day). In the aerobic biodegradation database, zero-order rate constants are reported where the author has determined this value.

A more accurate but time consuming approach would have been to plot the substrate concentration versus time. A straight line would signify zero-order kinetics and an exponential curve (or a straight line on a log linear paper) would indicate first-order kinetics. Priority was given to the determination of a first-order rate constant as many environmental models require the input of a first-order rate constant. This may not be strictly correct in all situations, such as when the substrate is present at high concentrations (above K_m), when substrate concentrations are toxic to the microbial population, when another substrate(s) is limiting the biodegradation rate or when the microbial population is significantly increasing or decreasing in size.

Recently, the common use of first-order rate constant values to describe the kinetics of biodegradation loss in natural systems has been criticized. Bekins et. al. suggest that the automatic use of firstorder kinetics without first determining whether the substrate concentration is less than the halfsaturation constant, K_m, is incorrect and can lead to substantial miscalculations of the biodegradation rate of a studied compound. Using first-order kinetics where the substrate concentration is higher than K_m will lead to an overprediction of the biodegradation rate, whereas in circumstances where the substrate concentration is much lower, the biodegradation rate is expected to be underestimated.

Buscheck et. al. state that first-order rate constants are generally appropriate for soluble plumes where a contaminant concentration of less than 1 mg/L is present. However, a database of LUFT sites in California reports that benzene concentrations are greater than 2 mg/L at 625 of 1092 sites. This suggests that concentrations at spill sites may often exceed the K_m value and that first-order kinetics may not adequately represent the biodegradation of the studied compound. First-order rate constants are, however, commonly used to describe kinetics in natural systems often because of the lack of sufficient data points and the ease with which these values can be calculated. Salanitro reports that several studies where BTEX concentrations range from <1 to 5000 ppb are adequately described by first-order kinetics.

First-Order Rate Constants

First-order rate constants are used as a convenient approximation of the kinetics of degradation of test substrates where there is no growth of the microbial population and a low concentration of the test substrate is present. Under these circumstances, the substrate concentration is lower than K_m and, over time, both the concentration of substrate and rate of degradation drop in proportion with each other. Thus, unlike zero-order kinetics, the rate of biodegradation in a first-order reaction is dependent on the substrate concentration and is represented by the differential:

$$\frac{dS}{dt} = k_1 S \tag{3}$$

and the integral:

$$k_1 = \frac{\ln \frac{S_0}{S}}{t} \tag{4}$$

where S_0=initial substrate concentration, S=substrate concentration at time=t, and k_1=the first-order rate constant.

During first-order rate reactions, the loss of substrate is exponential and follows a logarithmic curve. The rate constant is used to correlate the rate of the reaction with time. In a first-order reaction, a constant percent of the substrate is lost with time and the rate is described by either percent per time or the half-life. The half-life is easily visualized and is more commonly used. In contrast, a zero-order rate constant by definition equals the rate and is given in units of concentration/time. This is because the rate is linear and loss is constant with time.

Mineralization Rate Constants and Primary Biodegradation Rate Constants

Many experiments summarized in the aerobic

biodegradation database measured mineralization, defined as the complete biodegradation of a compound to carbon dioxide (CO_2) and water, in contrast to experiments which measure primary biodegradation, defined as the loss of the parent compound. These values are labeled as such in the rate constant comment field. Mineralization rate constants, in general, represent a minimum estimate of the total degradation of the compound; some of the CO_2 produced during the degradation of the original compound can be assimilated into natural, often high molecular weight compounds and is then generally not measured. In addition, once produced, CO_2 can be bound as carbonate within the study system. Thus, it is expected that unless degradation proceeds rapidly and completely to CO_2 and water, that mineralization rate constant values will be less than those measured for primary biodegradation.

Rate constants were collected from eight types of studies: laboratory column, field, groundwater grab sample, groundwater inoculum, in situ microcosm, lysimeter, reactor systems, and laboratory microcosm studies. The majority of studies summarized in the aerobic biodegradation database were laboratory microcosm studies. Laboratory microcosm studies can be further subdivided by the type of grab sample used: soil, sediment, surface water (including freshwater, estuarine, and seawater), and aquifer sediment and groundwater mixtures.

The information obtained from each of these studies ranged from published first-order rate constants to simply an indication or contraindication of biodegradation. In some cases, insufficient data were available to assess whether biodegradation had occurred; for these studies, the rate constant field was left blank. To ensure that loss of a contaminant was due to biodegradation and not just to abiotic or transport processes, an appropriate control was necessary to correct the data set. This can be a problem in laboratory studies that are incubated for a long period of time.

Mercuric chloride is known to adsorb to the clay component of soil or aquifer sediment reducing its efficacy whereas sodium azide only inhibits bacteria containing cytochromes. In addition, autoclaving may not be totally suitable, probably due to incomplete sterilization.

A control was used in laboratory studies to correct for non-biodegradation processes such as sorption to sediment or the glass jar, headspace volatilization, etc. Data from laboratory studies (column, groundwater grab sample, groundwater inoculum, reactor systems, and laboratory microcosm studies) were obtained from graphs or tables giving concentrations of the compound of interest at specific timepoints.

Lag periods were observed at times which is usually attributed to the need for acclimation. The initial microbial species present, their relative numbers, metabolic state and ability to acclimate once exposed to a chemical are likely to vary considerably depending upon environmental parameters such as temperature, conductivity, pH, oxygen concentration, redox potential, concentration, the presence/ absence of electron acceptors and donors, and effects, both synergistic and antagonistic, of associated microflora.

Lag periods were established either from the discussion in the paper or from looking at the data, and an appropriate initial and final concentration was chosen. The value used for the initial concentration was the concentration present following the lag period; therefore, all rate calculations for this project are independent of the associated lag period. Where a value of "0 Og/L" was reached as a final timepoint, an earlier time was chosen for the kinetics calculation, if possible; the use of zero as a denominator in the first-order rate equation would result in an "infinite" value.

If the concentration reached a value other than zero but leveled off at that point for the remainder of the experiment, the final concentration and time were chosen at the point where the concentration leveled off. In column studies, the time field in the database contains the retention time for the column, which is the value (Ot) used to calculate the rate constant; column experiments were usually run for long periods of time, which would allow for the development of an acclimated microbial population. The initial and final concentrations of the control within the chosen time period were obtained and the experimental data corrected for the loss shown by the control using the following equation:

$$C_{f,corr} = C_f \frac{Z_i}{Z_f} \tag{5}$$

where:

$C_{f,corr}$ = corrected final concentration of the contaminant (corrected for nonbiodegradation loss

C_f = final contaminant concentration, uncorrected

Z_i = initial control concentration

Z_f = final control concentration.

A first-order rate constant was then calculated for laboratory data using the corrected final contaminant concentration as follows:

$$k_1 = \frac{\ln \frac{C_i}{C_{f,corr.}}}{\Delta t} \tag{6}$$

where:

C_i = initial contaminant concentration

$C_{f,corr}$ = corrected final concentration of the contaminant (corrected for non-biodegradation loss)

Ot = time interval

k_1 = first-order rate constant.

In situ microcosms were designed to isolate a portion of the aquifer in order to make measurements directly in the field. This device is essentially a pipe divided into a test chamber and an equipment chamber, with two screens that permit water to be pumped both into and out of the interior of the pipe. Groundwater is pumped to the surface, spiked with the compounds of interest plus other nutrients and/or electron acceptors if wanted, and then reinjected. Because the test zone is isolated from the main aquifer, advective and dispersive processes are not important to the study results. Often, this method is used to give very specific results for a particular redox regime within an aquifer.

The data obtained from this type of study was similar to that for a laboratory microcosm where loss of substrate is monitored with time; rate constants were calculated using the same method as for the laboratory studies. In general, the field studies reported in this database are for aquifer environments. Only a limited number of aerobic aquifer studies were located, mainly because the oxygen initially present in groundwater will be rapidly used during oxidative degradation. This results in anaerobic conditions close to the source and within the contaminant plume. However, biodegradation data were reported for a few aerobic aquifer environments.

Data from field studies were generally reported for 1) plume studies where monitoring wells were placed along the centerline of a contaminant plume or for 2) continuous injection experiments where monitoring wells were placed in fences along the flow path fairly close to the injection point (often 2 and 5 meters away). Loss of a contaminant over distance does not necessarily indicate that the compound has undergone biodegradation. Significant loss in concentration along a flow path is often reported for compounds simply due to non-biological processes such as advection, dispersion, sorption, and dilution.

However, degradation is the only mechanism which leads to an actual loss of the contaminant. The most convenient way to correct for non-biodegradation processes in both plume and injection studies is to use compounds present in the contaminant plume or injection mixture that are 1) biologically recalcitrant and 2) have similar properties, such as Henry's Law constant and soil sorption coefficient, as the contaminant of interest. Most studies summarized in the aerobic biodegradation database used either chloride or bromide ion as a tracer satisfying the first condition of biological recalcitrance.

The procedure for obtaining a normalized data set from an aquifer field study was taken from Wiedemeier et. al. and is again outlined in the report accompanying the anaerobic biodegradation database. This method requires measurements of tracer and contaminant concentrations from a minimum of two points along a flow path in order to correct for the loss of the compound of interest due to transport processes. A mass balance approach has also been used by some researchers to determine the rate of biodegradation of specific contaminants in groundwater during a field study.

Mass flux of the studied contaminant through a line/ cluster of wells (a transect) is recorded instead of monitoring loss of the contaminant at specific points down the middle of a plume, as is typical for a plume centerline study. Wiedemeier et. al., suggests that the calculations involved are approximate and that often many of the required parameters necessary for the modeling are not available.

Biodegradation of organic compounds under aerobic conditions most often occurs when bacteria catalyze the breakdown of these molecules and then recover some of this chemical energy as ATP (adenosine triphosphate) which is absolutely necessary for maintenance of the bacterial cell. ATP is generated through a series of oxidation-reduction reactions (the electron transport chain) where electrons are sequentially transferred from one compound, the electron donor, to an electron acceptor.

The final or terminal electron acceptor in aerobic respiration is oxygen. Dissolved oxygen concentrations of 1 mg/L or greater are considered to define aerobic conditions. During aerobic respiration, the oxygen present in the environment is converted to water and thus the dissolved oxygen content can decrease. This is particularly significant in closed systems, as in a confined aquifer, where conditions can quickly become anaerobic with the metabolism of high concentrations of organic chemicals. Thermodynamically, the reduction of molecular oxygen to water is very favorable for the participating microorganisms. Because hydrocarbons are generally chemically reduced and stable, this is a preferred pathway over other redox pathways such as anaerobic chemical reduction.

Aerobic biodegradation results in the oxidation of the original compound. Metabolism of aliphatic compounds generally proceeds initially by production of the alcohol and then oxidation to the carboxylic acid which is susceptible to betaoxidation. In pure culture studies, aromatic hydrocarbons have been shown to biodegrade generally with the addition of one molecule of oxygen giving the dihydrodiol intermediate, usually with a cisstereochemistry. This intermediate is then oxidized forming the catechol which then allows for orthoor meta-cleavage of the aromatic ring structure.

Groundwater grab samples were considered separately as it has been shown that a large majority of microorganisms responsible for biodegradation in the subsurface environment are associated with the aquifer sediment surface. Therefore, rates collected during groundwater grab studies may not be as rapid as those where aquifer sediment is included. Laboratory microcosm studies are believed to give very good evidence of biodegradation at a specific location and can provide an "absolute mass balance" on a particular contaminant. In addition, the formation and measurement of metabolites can definitively show the biodegradation of the contaminant of interest.

However, results from a laboratory microcosm can be greatly influenced by many factors such as the source, collection, and condition of the grab sample (e.g. what is a representative source of material for that site?), the ratio of soil/sediment/aquifer sediment to water used in the microcosm, the type of sampling (repetitive or sacrificed), incubation conditions (e.g. substrate concentration, temperature), and the length of the study period (and its effect particularly on the initial microbial population during a long study period). If a sufficient supply of oxygen is not available to completely biodegrade the added organic compound/s then anaerobic conditions may prevail.

The mixing of a natural sample during its collection or the construction of a microcosm may also result in a "disturbance artifact" which is seen as an increase in the microbial activity of the sample. However, the influence of transport processes such as volatilization and adsorption cannot be measured in a microcosm experiment. If consideration of these processes is important, then field studies can be used to provide environmentally relevant data for a specific site, essentially showing whether the compound of interest can or cannot be biodegraded at that location.

The results for each compound are presented in the following sections. Separation of the data into mineralization and primary degradation studies was initially completed and each category was considered separately. A range was given to represent the dispersion of the data within the group as well as a median value, representing the central tendency of the data. In addition, frequency distribution histograms for the two types of studies are given for each compound with sufficient data. Within the subcategories of mineralization and primary degradation, each study was given equal weighting despite differences in how the study was carried out. Rate constants which were given as zero-order and could not be converted to first-order rate constants were not included in the statistical analysis.

BTEX Compounds

The BTEX group is composed of the water-soluble and monoaromatic compounds benzene, toluene, ethylbenzene, o-xylene, m-xylene, and p-xylene. In both laboratory and field studies, the biodegradation of all the BTEX compounds has been shown under aerobic conditions. There is a stoichometric requirement of 3 ppm O_2 to 1 ppm BTEX for the aerobic degradation of fuel hydrocarbons with rates of biodegradation appearing to slow for dissolved oxygen concentrations below 1 to 2 ppm in microcosm and field studies and below 1 ppm for soil column studies. Laboratory studies where 8 mg/L dissolved oxygen is initially present have been shown to rapidly biodegrade 2 mg/L or less of a BTEX mixture or a particular BTEX compound.

The majority of studies located for the BTEX compounds were for aquifer environments. Many aquifers become anaerobic during contaminant biodegradation due to the use of oxygen in aerobic respiration. Replacement of this oxygen from upgradient of the source, plume edges, infiltration of precipitation, or from vadose or saturated zone recharge is slower than its use during aerobic metabolism. Thus, the concentration of oxygen often becomes the rate-limiting factor in the biodegradation of the BTEX compounds in aquifer environments.

During laboratory studies this can be controlled by the addition of oxygen or hydrogen peroxide. Extrapolation of laboratory rate constants to field environments which are confined or semi-confined systems, however, should be made with caution both because oxygen conditions cannot be controlled as readily and because laboratory rates are often higher (up to an order of magnitude) than those reported from field experiments.

Benzene

While benzene is considered recalcitrant under anaerobic conditions, most evidence currently available shows that this compound is moderately degradable in the presence of oxygen. Degradation is thought to proceed via catechol to CO_2. 3.08 mg of oxygen are necessary to biodegrade 1 mg of benzene to CO_2 and water. This calculation does not include the energy requirement for cell maintenance and thus is not a conservative value. However, the value of 3.1 mg oxygen to degrade 1 mg benzene is suggested as a conservative estimate. Most of the located data for benzene under aerobic conditions were for aquifer environments. Field studies at six different locations consistently reported the biodegradation of benzene, giving half-life values ranging from 58 to 693 days.

The longer half-life was associated with an uncontaminated aquifer study. Initial concentrations of up to 25 mg/L were biodegraded under field conditions. Biodegradation of benzene was observed as well during in situ microcosm studies at two locations. Half-lives ranged from 1.4 to 103 days with an average half-life of 4 days. The high half-life value represents biodegradation in the groundwater only section of the in situ microcosm; half-life values obtained in the aquifer sediment + groundwater section were significantly lower. By far the most common type of study used to observe the biodegradation of benzene under aerobic conditions is the laboratory microcosm.

Mineralization half-lives for benzene in lab microcosm studies ranged from 7 to 1195 days with the high value representing a study from an uncontaminated site. Microcosms established with sediment from a contaminated and a biostimulated region in the aquifer, measured during the same study, showed more rapid mineralization rates. The average half-life for mineralization was 53 days. In comparison, microcosm studies measuring primary biodegradation reported half-lives ranging from 0.2 to 679 days with an average value of 1.5 days. Initial concentrations of up to 50 mg/L were reported in these experiments without obvious deleterious effect.

In general, however, initial concentrations of 5 mg/L or less were utilized. No biodegradation was reported for four lab microcosm studies. A study by the American Petroleum Institute, 1994A, reports that benzene was not biodegraded in the presence of 85% methanol over 278 days. This result was not unexpected as sufficient oxygen was available to degrade only 5% of the initially added methanol. This suggests that anaerobic conditions may have occurred rapidly within this microcosm. Benzene in the presence of 300 mg/L ethanol was not biodegraded over a period of 3 days.

However, aquifer material collected from the same site and incubated without ethanol and either with or without nutrients was able to rapidly degrade benzene. Pristine aquifer sediment was unable to biodegrade benzene over a 7-day period; however, when creosote-contaminated aquifer sediment was used, greater than 31% of the initially added benzene was biodegraded over 7 days. While benzene was degraded in groundwater alone and in river water during a study by Vaishnav and Babeu, it was not biodegraded in the presence of harbor water collected in Lake Superior.

The addition of both nutrients and an enriched microbial culture isolated from sewage resulted in the biodegradation of this compound indicating that bacteria capable of biodegrading benzene were either not present or not present in sufficient numbers to significantly remove benzene in the natural harbor water over a 20-day period. Laboratory column experiments by Anid et. al. and Alvarez et. al. report that benzene was not biodegraded under certain circumstances. Anid et. al. reported that columns supplemented with hydrogen peroxide but not columns supplemented with nitrate were able to degrade benzene.

The nitrateamended columns may have exhibited nitrate-reducing conditions as over 60 mg/L BTEX mixture was initially added. However, no attempt was made by the authors to distinguish through end product measurements whether conditions remained aerobic or became nitrate-reducing. Alvarez et. al. showed biodegradation of benzene in laboratory columns fed with acetate and benzoate as cosubstrates. However, preacclimated sediment exposed to acetate and sediment columns which received no preacclimation period were unable to biodegrade benzene while a column which had been preacclimated to benzoate readily biodegraded this column. Benzene is expected to biodegrade fairly readily under most aerobic environmental conditions.

Toluene

Biodegradation of toluene is expected to occur readily in aerobic environments. Degradation is believed to proceed via 3-methylcatechol to CO_2. 3.13 mg of oxygen are necessary to biodegrade 1 mg of toluene to CO_2 and water. This calculation does not include the energy requirement for cell maintenance and thus is not a conservative value. Most of the located data for toluene under aerobic conditions were for aquifer or soil environments. Aquifer field studies at two different locations reported the biodegradation of toluene giving half-life values ranging from 1.3 to 77 days.

The longer half-life was associated with an uncontaminated aquifer study. Rate constants measured during a stream field study were very high giving half-lives less than 1 day in every case. Biodegradation of toluene was observed as well during several in situ microcosm studies at Vejen City, Denmark. Half-lives ranged from 1.7 to 103 days with an average half-life of 4.5 days. The high half-life value represents biodegradation in the groundwater only section of the in situ microcosm; half-life values reached in the aquifer sediment + groundwater section were significantly lower. By far the most common type of study used to observe the biodegradation of toluene under aerobic conditions is the laboratory microcosm.

Mineralization half-lives for toluene in lab microcosm studies ranged from 5 to 693 days. The average half-life for mineralization was 36 days. In comparison, microcosm studies measuring primary biodegradation reported half-lives ranging from <1 to 495 days with an average value of 0.5 days. No biodegradation was reported for nine lab microcosm studies. A study by the American Petroleum Institute, 1994A, reports that toluene was not biodegraded in the presence of 85% methanol over 278 days. This result was not unexpected as sufficient oxygen was available to degrade only 5% of the initially added methanol. This suggests that anaerobic conditions may have occurred rapidly within this microcosm.

In addition, the bacteria may have preferentially degraded the structurally simpler methanol or the concentration of methanol added may have been toxic to the indigenous bacterial population. Davis and Madsen, 1996, reported the biodegradation of toluene in three different soils under varying moisture and concentration levels. No biodegradation of toluene was reported in air-dried soils over 30 days; however, toluene added to the same soil, when 100% moisture was present, was biodegraded with a half-life of less than 1 day.

Some researchers discoverd that toluene was not biodegraded in soil which had been incubated at a moisture level of 2.5%; however, half-lives of 2 and 71 days were measured for moisture levels of >16 and 5%, respectively. Initial concentrations of up to 50 mg/L were reported in these experiments without obvious deleterious effect. In general, however, initial concentrations of 5 mg/L or less were utilized. Concentrations of 500 mg/kg soil were shown to be inhibitory to the biodegradation of toluene while at concentrations of 50 mg/kg soil or less, half-lives of less than 1 day were again measured.

Degradation of toluene may also be limited by concentrations of inorganic nutrients in its immediate environment. Allen-King et. al. reports that nitrogen limitation caused the degradation rate of toluene to be almost immeasurably slow in uncontaminated soil grab samples without added nutrients. However, when nitrogen was added, a half-life of 4.5 days was obtained for this compound in the same soil. Pristine aquifer sediment was unable to biodegrade benzene over a 7 day period; however, when creosote-contaminated aquifer sediment was used, greater than 18-22% of the initially added toluene was biodegraded over 7 days.

Ethylbenzene

Ethylbenzene is moderately degradable in the presence of oxygen. Degradation is thought to proceed via 3-ethylcatechol to CO_2. 3.17 mg of oxygen are necessary to biodegrade 1 mg of ethylbenzene to CO_2 and water. As for the other BTEX compounds, most of the located data for ethylbenzene under aerobic conditions were from aquifer environments. Field studies at two different locations reported the biodegradation of ethylbenzene. Half-life values ranged from 87 to 231 days for an uncontaminated aquifer study. No in situ microcosm studies were found for this compound. Only one study was found which measured the rate of mineralization for ethylbenzene.

A mineralization half-life of 19 days was calculated in river water. By far the most common type of study used to observe the biodegradation of ethylbenzene under aerobic conditions is the laboratory microcosm. Half-lives ranging from 1 to 231 days with an average value of 4.4 days were measured. The high half-life reported by the American Petroleum Institute represents a microcosm study where 85% methanol was present. When methanol was not present, a half-life of 22 days was obtained. Initial concentrations of up to 20.6 mg/L were reported in these experiments without obvious deleterious effect. In general, however, initial concentrations of less than 2 mg/L were utilized.

o-Xylene

o-Xylene is moderately degradable in the presence of oxygen. Degradation is thought to proceed via 3,4-dimethylcatechol to CO_2. 3.17 mg of oxygen are necessary to biodegrade 1 mg of o-xylene to CO_2 and water. As for the other BTEX compounds, most of the located data for o-xylene under aerobic conditions were for aquifer environments. Two different field studies from the same location reported the biodegradation of o-xylene giving half-life values ranging from 116 to 173 days. Biodegradation of o-xylene was observed as well during in situ microcosm studies at Vejen City, Denmark. Half-lives ranged from 7 to 408 days days with an average half-life of 18 days.

The high half-life value represents biodegradation in the groundwater only section of the in situ microcosm; half-life values obtained in the aquifer sediment + groundwater section were significantly lower. By far the most common type of study used to observe the biodegradation of o-xylene under aerobic conditions is the laboratory microcosm. Half-lives ranging from 1 to 41 days with an average value of 4.3 days were measured. Initial concentrations of up to 20 mg/L were reported in these experiments without obvious deleterious effect. In general, however, initial concentrations of less than 2 mg/L were utilized.

No biodegradation was reported for several microcosm studies. A study by the American Petroleum Institute reports that o-xylene, as well as benzene, toluene and both meta- and para-xylene, were not biodegraded in the presence of 85% methanol over 278 days. This result was not unexpected as sufficient oxygen was available to degrade only 5% of the initially added methanol (over 7 grams were initially added). This suggests that anaerobic conditions may have occurred rapidly within this microcosm. In addition, the bacteria may have preferentially degraded the structurally simpler methanol or the concentration of methanol added may have been toxic to the indigenous bacterial population.

Pristine aquifer sediment was unable to biodegrade o-xylene over a 7 day period; however, when creosote-contaminated aquifer sediment was used, greater than 19-21% of the initially added o-xylene was biodegraded over 7 days. Thomas et. al. reported that o-xylene was readily biodegraded in uncontaminated aquifer material with a half-life of 7 days but when either contaminated or biostimulated aquifer material was used instead, no biodegradation was reported over 21 days. Laboratory microcosms which contained double concentrations of nitrogen and phosphorus nutrients were able to significantly biodegrade a xylene mixture; however, unamended and single nutrient amendments to the microcosm were insufficient to encourage biodegradation of this mixture.

Laboratory column experiments by Broholm et. al. and Alvarez et. al. report that oxylene was not biodegraded under certain circumstances. Alvarez et. al. showed biodegradation of o-xylene in laboratory columns fed with acetate and benzoate as cosubstrates. However, preacclimated sediment exposed to acetate and columns which received no preacclimation period were unable to biodegrade o-xylene while a column which had been preacclimated to benzoate readily biodegraded o-xylene. Broholm et. al. were not able to explain their results showing no to slight biodegradation of o-xylene over an experimental period of 3.5 days.

It may be that this time period was not sufficient for acclimation of the bacterial population within the clay column. The median for the primary biodegradation rate constant of o-xylene, considering all studies, is 0.054/day (N = 92); a range of not biodegraded to 7.625 is reported. o-Xylene is expected to biodegrade readily under most aerobic environmental conditions.

m-Xylene

m-Xylene is moderately degradable in the presence of oxygen. Degradation is thought to proceed via 2,4-dimethylcatechol to CO_2. 3.13 mg of oxygen are necessary to biodegrade 1 mg of m-xylene to CO_2 and water. As for the other BTEX compounds, most of the located data for m-xylene under aerobic conditions were for aquifer environments. Two different field studies from the same location reported the biodegradation of m-xylene giving half-life values ranging from 50 to 77 days. No in situ microcosm studies were found for this compound. By far the most common type of study used to observe the biodegradation of m-xylene under aerobic conditions is the laboratory microcosm.

Half-lives ranging from 1 to 27 days with an average value of 4.3 days were measured. Initial concentrations of up to 20 mg/L were reported in these experiments without obvious deleterious effect. In general, however, initial concentrations of less than 2 mg/L were utilized. No biodegradation was reported for several microcosm studies.

A study by the American Petroleum Institute, 1994A, reports that m-xylene was not biodegraded in the presence of 85% methanol over 278 days. This result was not unexpected as sufficient oxygen was available to degrade only 5% of the initially added methanol. In addition, the bacteria may have preferentially degraded the structurally simpler methanol or the concentration of methanol added may have been toxic to the indigenous bacterial population.

Laboratory microcosms which contained double concentrations of nitrogen and phosphorus nutrients were able to significantly biodegrade a xylene mixture; however, unamended and single nutrient amendments to the microcosm were insufficient to encourage biodegradation of this mixture. Pristine aquifer sediment was unable to biodegrade m-xylene over a 7-day period; however, when creosote-contaminated aquifer sediment was used, greater than 19-22% of the initially added m-xylene was biodegraded over 7 days.

The nitrate-amended columns may have exhibited nitrate-reducing conditions as over 60 mg/L BTEX mixture was added. However, no attempt was made by the authors to distinguish through end product measurements whether conditions remained aerobic or became nitrate-reducing. The median for the primary biodegradation rate constant of m-xylene is 0.057/day (N = 32); a range from not biodegraded to 0.76/day is reported. m-Xylene is expected to biodegrade fairly readily under most aerobic environmental conditions.

p-Xylene

p-Xylene is moderately degradable in the presence of oxygen. Degradation is thought to proceed via 3,6-dimethylcatechol to CO_2. 3.13 mg of oxygen are necessary to biodegrade 1 mg of p-xylene to CO_2 and water. As for the other BTEX compounds, most of the located data for p-xylene under aerobic conditions were for aquifer environments. Three field studies from two different locations reported the biodegradation of p-xylene giving half-life values ranging from 49 to 231 days. No in situ microcosm studies were found for this compound. By far the most common type of study used to observe the biodegradation of p-xylene under aerobic conditions is the laboratory microcosm.

Half-lives ranging from 1 to 28 days with an average value of 4.4 days were reported. Initial concentrations of up to 20 mg/L were reported in these experiments without obvious deleterious effect. In general, however, initial concentrations of less than 2 mg/L were utilized. No biodegradation was reported for two microcosm studies. A study by the American Petroleum Institute, 1994A, reports that p-xylene was not biodegraded in the presence of 85% methanol over 278 days. This result was not unexpected as sufficient oxygen was available to degrade only 5% of the initially added methanol.

In addition, the bacteria may have preferentially degraded the structurally simpler methanol or the concentration of methanol added may have been toxic to the indigenous bacterial population. Laboratory microcosms which contained double concentrations of nitrogen and phosphorus nutrients were able to significantly biodegrade a xylene mixture; however, unamended and single nutrient amendments to the microcosm were insufficient to encourage biodegradation of this mixture. Anid et. al. reports that columns supplemented with hydrogen peroxide but not columns supplemented with nitrate were able to degrade p-xylene.

The nitrateamended columns may have exhibited nitrate-reducing conditions as over 60 mg/L BTEX mixture was added. However, no attempt was made by the authors to distinguish through end product measurements whether conditions remained aerobic or became nitrate-reducing. The median for the primary biodegradation rate constant of p-xylene, considering all studies, is 0.052/day (N = 27); a range of not biodegraded to 0.56/day is reported. p-Xylene is expected to biodegrade fairly readily under most aerobic environmental conditions.

Polycyclic Aromatic Hydrocarbon (PAH) Compounds

PAH compounds are ubiquitous compounds in the environment due to their production during incomplete combustion and pyrolytic processes. In addition, effluents from wood-preserving (creosoting sites), refinery and oil waste, waste incineration, coal gasification and liquification, and coal tar distillation sites often provide high concentrations of PAH compounds. Structurally, PAHs are considered to have two or more fused benzene ring structures in various arrangements. This database includes several PAHs including naphthalene (2 rings), fluorene (3 rings), benzo(a)anthracene, chrysene, fluoranthene, and pyrene (4 rings), and benzo(a)pyrene (5 rings). Based on pure culture studies, bacterial biodegradation of PAHs is believed to proceed via oxidation of the ring structure using dioxygenase enzymes in the presence of molecular oxygen.

The cis-dihydrodiol is then dehydrogenated forming the catechol which leaves the aromatic ring open to enzymatic cleavage by another dioxygenase activity. The two hydroxyl groups must be located either ortho or para to each other for ring cleavage to occur. The lower molecular weight PAH compounds are generally more rapidly degraded than the higher molecular weight PAH compounds. This was shown by Park et. al., in a study investigating the biodegradation of 14 PAH compounds in two different soils.

Compounds with two or three rings were extensively degraded while, in general, compounds with four or five rings were resistant to biodegradation. Sorption of the PAH compound to the soil or sediment appears to strongly affect its ability to be biodegraded. Maliszewska-Kordybach examined the physical properties and their correlation to the biodegradation rate reported for four PAH compounds in ten different soils varying in pH and organic matter content. The length of time the PAH was in contact with the soil had a strong effect on the importance of the physical property/properties relationship to persistence.

Initially, volatilization, reflected by Henry's Law constant was correlated to loss of the PAH in soil. However, over time, other properties such as water solubility and sorption became more important in determining the resistance of a particular PAH to degradation. Lower water solubilities were correlated to lower biodegradation rates. Sorption properties, however, showed the strongest influence in determining PAH persistence in soil. Loehr and Webster reviewed the current literature examining the effect exposure time had on the biodegradation in soil of compounds ranging from chlorinated aliphatics to PAHs.

They proposed that studies using soil freshly amended with a compound cannot fairly reflect the biodegradative ability of a compound which has been in contact with soil for a long period of time and has thus undergone weathering processes. Sorption of compounds mainly by soil organic matter has been indicated, with increasing organic matter generally resulting in increasing sorption. It is believed that a compound in contact with soil for a long time will be more strongly sorbed, have a lesser ability to desorb and will be generally less bioavailable that a compound that was recently spiked into the same soil environment. This is reflected experimentally by an initial period of relatively rapid biodegradation followed by a phase of little to no further loss.

Keck et. al. looked at PAH disappearance as the single compound, in a defined mixture of PAH compounds, in oil refinery waste, and in a creosote wood-preserving waste. The degradation of 3-ring PAH compounds was similar in all four systems; however, PAH compounds with 4 or more rings were more rapidly degraded in the oil-refinery and the wood-preserving waste. This suggested to the authors that co-oxidation of these larger PAH compounds was occurring. Rates of biodegradation also appear to be higher in environments where the medium is contaminated when compared to sites which are initially pristine.

Naphthalene

Naphthalene is a PAH with a 2-ring structure. In general, PAHs that contain three or fewer rings tend to be biodegraded more rapidly than compounds containing four or five benzene rings. First-order rate constants for naphthalene ranged from 0.0064/day to 5.0/day, with the majority of the values occurring between 0.1 to 1.0/day. These values correspond to half-lives of <1 to 108 days. The highest rate constant (5.0/day) resulted from polluted groundwater obtained downgradient from a landfill while the lowest rate constant was calculated from a field experiment in which naphthalene was injected into a series of wells and its concentration was measured over time.

The primary biodegradation of naphthalene resulted in the following products: cis-1,2-dihydroxy-1,2-dihydronaphthalene, 1-naphthol, salicylic acid and catechol. Naphthalene will be degraded rapidly by indigenous microorganisms in soils and groundwater that have previously been exposed to PAHs. The rate constant determined from sediment in an oil polluted stream was 3.36/day while the rate constant determined from sediment in a non-polluted stream was less than 0.0096/day.

The median for the primary biodegradation rate constant of naphthalene is 0.308/day (N = 49); a range of 0 to 5.0/day is reported. The median for the mineralization rate constant of naphthalene is 0.023/day (N = 47); a range of 0 to 3.34/day is reported.

Fluorene

Fluorene is a PAH with a 3-ring structure, that is virtually insoluble in water. In general, PAHs that contain three or fewer rings tend to be biodegraded more rapidly than compounds containing four or five benzene rings. First-order rate constants were in the range of 0.0018/day to 0.33/day, corresponding to half-lives of about 2 to 385 days. The smallest rate constant was observed in a pristine soil with no history of pollution, while the largest rate constant was observed in creosote-contaminated aquifer material obtained from a wood preserving site.

One study reported 9-fluorenone as a primary biodegradation product of fluorene. The median for the primary biodegradation rate constant of fluorene is 0.015/day (N = 28); a range of 0.0018 to 0.33/day is reported. The median for the mineralization rate constant of fluorene is 0.019/day (N = 7); a range of 0.00385 to 0.05/day is reported. For the most part, fluorene is expected to biodegrade slowly under aerobic conditions in the environment.

Benzo(a)anthracene

Benzo(a)anthracene is a PAH with a 4-ring structure, that is virtually insoluble in water. In general, PAHs that contain more than 3 rings tend to be biodegraded father slowly in the environment. Firstorder rate constants were in the range of 0.000033/day to 0.116/day, corresponding to half-lives of about 6 to 21,000 days. The majority of the rate constants were less than 0.01/day. There were also several studies for which no rate constant was reported or could be calculated since the compound was not observed to biodegrade during the course of the study. This included an experiment where no measurable biodegradation was observed in a coal-tar polluted soil over a 80 day incubation period.

Similar results were obtained from other studies with shorter incubation periods. Biodegradation reaction products were reported as benzo(a)anthracene-7,12-dione and unspecified phenol and quinone metabolites (Hinga et. al., 1980). The median for the primary biodegradation rate constant of benzo(a)anthracene, considering all studies, is 0.0035/day (N = 27); a range of 0.000096 to 0.072/day is reported. The median for the mineralization rate constant of benzo(a)anthracene is 0.0029/day (N = 32); a range of not biodegraded to 0.116/day is reported. For the most part, benzo(a)anthracene is expected to biodegrade slowly under aerobic conditions or be recalcitrant in the environment.

Chrysene

Chrysene is a PAH with a 4-ring structure, that is virtually insoluble in water. In general, PAHs that contain more than 3 rings tend to be biodegraded rather slowly in the environment. First-order rate constants were in the range of 0.0005/day to 0.037/day, corresponding to half-lives of about 19 to 1,400 days. The majority of the rate constants were smaller than 0.01/day. There were also several studies for which no rate constant was reported or could be calculated since the compound was not observed to biodegrade during the course of the study. This included an experiment where no measurable biodegradation was observed in an uncontaminated soil over a 56-day incubation period.

Similar results were obtained from other studies with shorter incubation periods. No reaction products were reported for the degradation of chrysene. The median for the primary biodegradation rate constant of chrysene is 0.003/day (N = 31); a range of not biodegraded to 0.037/ day is reported. The median for the mineralization rate constant of chrysene is 0.0037/day (N = 20); a range from not biodegraded to 0.035/day is reported. Chrysene is expected to biodegrade slowly under aerobic conditions or to be recalcitrant in the environment.

Fluoranthene

Fluoranthene is a PAH with a 4-ring structure. In general, PAHs that contain more than 3 rings tend to be biodegraded rather slowly in the environment. The primary biodegradation rate constant for fluoranthene ranges from 0.0015/day to 0.045/day, corresponding to half-lives of about 15 to 462 days. The lowest rate constant is derived from an unpolluted pristine soil, while the largest rate constant was observed in aquifer material from a heavily-polluted wood-preserving facility . The half-life of fluoranthene in pristine sediment obtained from New Zealand was reported as greater than 100 days, further illustrating that biodegradation will be extremely slow in non-acclimated environments.

No reaction products have been reported from the primary biodegradation of fluoranthene. The median for the primary biodegradation rate constant of fluoranthene is 0.0048/day (N = 20); a range of not biodegraded to 0.045/day is reported. No mineralization data are available. Fluoranthene is expected to biodegrade slowly or be recalcitrant under most environmental conditions.

Pyrene

Pyrene is a PAH with a 4-ring structure. In general, PAHs that contain more than 3 rings tend to be biodegraded rather slowly in the environment. First-order rate constants were in the range of 0.00036/day to 0.143/day, with the majority of the rate constant values less than 0.01/day. These values correspond to half-lives of 5 to 1,925 days. The smallest rate constant was observed in an unpolluted sandy loam that had a mixture of 16 other PAHs. This rate constant was derived from the reported half-life of 1,900 days. The largest rate constants (0.136/day and 0.143/day) were observed in unpolluted soil with roughly 80% moisture content. These two rate constants are much larger than those of any other study including those obtained from material at industrial locations with acclimated microbes.

There were also several studies for which no rate constant was reported or could be calculated since the compound was not observed to biodegrade over the course of the study. This included one experiment where no measurable biodegradation was observed in an unpolluted sandy loam during a 240-day incubation period. Similar results were obtained from other studies with shorter incubation periods. No reaction products were reported for the biodegradation of pyrene.

The median for the primary biodegradation rate constant of pyrene, considering all studies, is 0.00345/day (N = 40); a range of not biodegraded to 0.052day is reported. The median for the mineralization rate constant of pyrene is 0.006/day (N = 18); a range of not biodegraded to 0.143/day is reported. Pyrene is expected to biodegrade slowly under aerobic conditions or to be recalcitrant in the environment.

Benzo(a)pyrene

Benzo(a)pyrene is a PAH with a 5-ring structure that is virtually insoluble in water. In general, PAHs that contain more than 3 rings tend to be biodegraded rather slowly in the environment. Biodegradation of benzo(a)pyrene appears to occur at an extremely slow rate when compared to other PAHs. First-order rate constants ranged from 0.00006/day to 0.057/day. These values correspond to half-lives of 12 to 11,552 days. The smallest rate constant was observed in soil that contained 8 other PAHs. The biodegradation rate of benzo(a)pyrene was slower than the other compounds, indicating microbes may selectively degrade PAH compounds.

The largest rate constant was observed in creosote-contaminated groundwater obtained from a wood-preserving facility. There were also several studies for which no rate constant was reported or could be calculated since the compound was not observed to biodegrade over the course of the study. This included an experiment where no measurable biodegradation was observed in a polluted sandy loam over a 315-day incubation period. Similar results were obtained from other studies with shorter incubation periods. No reaction products were reported for the degradation of benzo(a)pyrene.

The median for the primary biodegradation rate constant of benzo(a)pyrene is 0.0027/day (N = 23); a range of not biodegraded to 0.057/day is reported. The median for the mineralization rate constant of benzo(a)pyrene is 0.00015/day (N = 25); a range of not biodegraded to 0.0037/day is reported. For the most part, benzo(a)pyrene is expected to be recalcitrant under most environmental conditions.

Chlorinated Aliphatic Compounds

While anaerobic biodegradation of highly chlorinated aliphatic compounds generally proceeds readily in strong reducing, anaerobic environments, aerobic degradation of these compounds is less likely. The degree of chlorination of the aliphatic compound will determine its ability to be aerobically biodegraded. Compounds such as tetrachloroethylene and carbon tetrachloride, which are highly oxidized and highly chlorinated, will not be readily biodegraded under aerobic conditions while less oxidized and less highly chlorinated compounds such as dichloromethane and vinyl chloride will.

The major route of degradation for these less chlorinated compounds is via a cometabolic pathway using broad specificity bacterial oxygenases. In this case, the chlorinated aliphatic is not used as a growth substrate but is fortuitously degraded during the biodegradation of other non-chlorinated compounds (or methane) present in the bacterial environment. Methanotrophs (methane-oxidizing bacteria) are able to metabolize some chlorinated aliphatics during a cometabolic process due to the methane monooxygenase enzyme complex.

However, the rate of aerobic degradation by methanotrophs also appears to lessen with an increasing number of chlorine atoms on the molecule. Tetrachloroethylene is apparently not susceptible to degradation by methanotrophs. In aerobic environments, as found in unsaturated soil or surface water, processes such as leaching or volatilization may be more important in the loss of highlychlorinated aliphatic compounds than biodegradation.

Tetrachloroethylene

This compound, while readily biodegraded under anaerobic conditions through reductive dechlorination processes, is resistant to biodegradation under aerobic conditions. Reductive dechlorination does not occur at oxygen concentrations greater than 0.5 mg/L. Enzien et. al. studied the degradation of tetrachloroethylene in laboratory columns where the measured dissolved oxygen concentration did not drop below 1.6 mg/L. Significant degradation of this compound occurred and the authors postulated that anaerobic microsites were responsible although the column was kept under bulk aerobic conditions. MPN counts of both aerobic and anaerobic bacteria showed that abundant populations of each existed in the aquifer sediment.

Methane was measured in the column indicating that methanogens were present and providing more evidence that anaerobic conditions, as microsites, may have been present. Finally, cis-dichloroethylene was reported as the major metabolite of tetrachloroethylene degradation, again suggesting that anaerobic biodegradation was responsible for the breakdown of this compound during the experiment.

A groundwater recharge field study by Roberts et. al. reporting the biodegradation of tetrachloroethylene, injected water with dissolved oxygen concentrations of 1.86 mg/L which may not have been sufficient to prevent anaerobic conditions from developing in the aquifer.

In both oxygen-saturated and methanotrophic lab microcosm conditions, tetrachloroethylene was minimally degraded from 0 to 6% over a 56-day period. No measurable oxygen consumption or carbon dioxide production was seen in these microcosms relative to the control microcosms; however, some loss of methane in the methanotrophic microcosms, in excess of that found in the controls, was reported suggesting possible unmeasurable transformation by a methanotrophic population. The remainder of the studies summarized in the database show that aerobic biodegradation of tetrachloroethylene does not occur even given time frames of over 700 days. Studies incorporating methane and oxygen in the microcosm atmosphere also reported that tetrachloroethylene was not biodegraded, again showing that this compound is not susceptible to methanotrophic cometabolism.

Dichloromethane

Several studies using pure cultures or mixed defined bacterial cultures show that dichloromethane, unlike the more highly chlorinated aliphatic compounds can be utilized by bacteria as a carbon and energy source for growth. Biodegradation of this compound is believed to be catalyzed initially by a dehalogenase resulting in the formation of formaldehyde and hydrochloric acid. Dichloromethane is expected to be moderately degradable under aerobic conditions. Davis and Madsen investigated the aerobic degradation of dichloromethane in several soils without the addition of methane.

The rate of degradation was dependent on the soil type, substrate concentration, organic carbon availability and redox condition of the particular soil studied. Rates were significantly higher for concentrations between 0.5 mg/L (T1/2 of 1.3 days) and 5 mg/L (T1/2 of 191 days) for the same soil although degradation was shown at concentrations up to 50 mg/L. Henson et. al. reported the biodegradation of dichloromethane following the addition of methane to stimulate the naturally occurring population of methanotrophs.

Without the prior addition of methane only 43% of the initially added dichloromethane was degraded in 6 days whereas the enrichment procedure, with methane added over the 6 weeks before the addition of dichloromethane, showed 98% loss over 6 days. The median for the primary biodegradation rate constant of dichloromethane, considering all studies, is 0.0546/day (N = 8); a range of 0.00362 to 0.533/day is reported. Dichloromethane is expected to be fairly readily biodegraded under most aerobic environmental conditions.

Phenol and Substituted Phenols

Both phenol and the cresol isomers are readily degraded under aerobic conditions in environmental media. The cresols are monosubstituted phenols commonly used in disinfectants and fumigants, in photographic developers, as a manufacturing intermediate in the production of synthetic resins and in explosives. Because of their common usage, these compounds are often found in the environment and are listed as priority pollutants by the U.S. EPA. Phenol is expected to biodegrade more rapidly than the cresol isomers. Phenols with ortho- or para- substituents were shown to be preferentially degraded over those with meta-substitution. m-Cresol is thought to exhibit a greater stability of the benzene ring making it more difficult to biodegrade than the other two isomers.

Phenol

Phenol is rapidly biodegraded in the environment under aerobic conditions, usually without any apparent lag time. First-order biodegradation rate constants were in the range of 0.006/day T1/2=116 days) to 11/day (T1/2<1 day), with the majority of the values greater than 1/day. The lowest rate constant was derived from unpolluted aquifer material 12-12.6 m below the soil surface and had a lag period of 16 days. Biodegradation of phenol in soils under aerobic conditions was also rapid. The median for the primary biodegradation rate constant of phenol, considering all studies, is 0.21/day (N = 38). The median for the mineralization rate constant of phenol, considering all studies, is 1.56/day (N = 26).

o-Cresol

o-Cresol is readily degraded under aerobic conditions. Biodegradation of this compound is believed to proceed via hydroxylation to 2-methylcatechol followed by ring cleavage. While no field studies were located for this compound, in situ microcosm studies conducted at Vejen City, Denmark showed that o-cresol was readily degraded in an aquifer environment with an average half-life of 2.6 days. Rate constants were mainly collected for laboratory microcosm studies with calculated half-lives ranging from less than 1 day to 10 days.

The average half-life for these data was less than 1 day. Rate constants reported for a study using groundwater as an inoculum for the degradation of o-cresol were considerably higher than those reported in other study types. The median for the primary biodegradation rate constant of o-cresol, considering all studies, is 0.4/day (N = 22); a range of 0.069 to 4.61/day is reported. The mean for the lab microcosm studies alone is 0.49/day. o-Cresol is expected to be readily biodegraded

m-Cresol

m-Cresol is rapidly degraded under aerobic conditions. Biodegradation is believed to proceed via hydroxylation to 3-methylcatechol followed by ring cleavage. No field or in situ microcosm studies were located for this compound. The average half-life for this data was approximately 1.5 days. The high half-life was taken from a study measuring the biodegradation of m-cresol at different horizons in a sediment core from a pristine aquifer site. The half-lives measured for this core varied from 28 to 198 days depending on the location of the sediment used in the laboratory microcosm.

A single mineralization study was reported giving a half-life of 714 days for m-cresol in uncontaminated aquifer sediment. Loehr and Matthews reported that biodegradation of m-cresol was considerably slower in an acid soil (pH 4.8) when compared to degradation in a basic soil (pH 7.8) giving half-lives of 0.6 and 11.3 days, respectively. This difference was not seen in the same magnitude for the other cresol isomers although degradation of the cresols in the acidic soil was slower in all cases. The median for the primary biodegradation rate constant of m-cresol, considering all studies, is 0.133/day (N = 19); a range of 0.0035 to 1.16/day is reported. m-Cresol is expected to be readily biodegraded under most aerobic environmental conditions.

p-Cresol

p-Cresol is readily degraded under aerobic conditions. Biodegradation is believed to proceed via hydroxylation to 4-methylcatechol followed by ring cleavage. No field or in situ microcosm studies were located for this compound. Rate constants were exclusively collected for laboratory microcosm studies aside from one groundwater grab study with calculated half-lives ranging from less than 1 day to 27 days. The average half-life for this data was less than 1 day. A single mineralization study was reported giving an approximate half-life of 10 days for p-cresol in aquifer sediment.

In a factorial experiment, it was found that the mineralization of p-cresol was unaffected by the presence of other compounds commonly found in creosote. The median for the primary biodegradation rate constant of p-cresol is 1.75/day (N = 28), a range of 0.079 to 13.15/day is reported. p-Cresol is expected to be readily biodegraded under most aerobic environmental conditions.

Ketones

Data were located for two of the three original members of this group, acetone, methyl ethyl ketone and methyl isobutyl ketone. The ketones, in general, are readily biodegraded in soil and water and thus are expected to naturally attenuate in the environment.

Acetone

Acetone is expected to biodegrade rapidly under aerobic conditions. Hwang et. al. reported that mineralization of acetone in freshwater samples, varying time of collection and concentration, followed multiphasic kinetics which was best described by the Michaelis-Menton model. First-order biodegradation rate constants of 0.025/day to 7.3/day were calculated from these data. The largest values were observed for low initial concentrations of acetone (0.5 μg/L), while the smallest values were observed for high initial concentrations (500 μg/L). Acetone was not biodegraded over 30 days in a non-amended microcosm using soil and groundwater from an contaminated solvent storage site.

Methyl ethyl ketone

Methyl ethyl ketone is expected to biodegrade rapidly

under aerobic conditions. Only two papers were located, however, reporting the degradation of this compound in environmental media. In pure culture experiments, methyl ethyl ketone was initially hydroxylated to acetoin, and then degraded to diacetyl and acetic acid. Dojlido reports that concentrations up to 800 mg/L were not toxic to microbial activity during typical screening tests using municipal wastewater. In river die-away studies, methyl ethyl ketone initially present at 20 mg/L was degraded within one to two days. A second addition of 20 mg/L methyl ethyl ketone to the same river water, once degradation from the first addition was complete, resulted in 100% degradation in less than one day.

Delfino and Miles reported that methyl ethyl ketone, at 1 mg/L, was completely biodegraded in an aerated, previously anaerobic groundwater within 12 days including a 4 to 5 day lag phase. A spill of methyl ethyl ketone, resulting in groundwater contamination, was remediated by air stripping and then by biological degradation. While there was insufficient information in the paper to report a rate constant, the authors stated that laboratory tests showed that the indigenous bacteria were capable of degrading this compound. The median for the primary biodegradation rate constant of methyl ethyl ketone, based on limited data, is 0.69/day (N = 3); a range of 0.4 to 1.4/day is reported.

Plumping for Biodegradation: A Guide to Septic System

The term "septic" is an adjective relating back to the word "sepsis," a noun describing the "toxic condition resulting from the spread of bacteria or their products from a source of infection." Today we are perhaps more familiar with "antiseptic," which of course is literally "opposing sepsis ..." and describes efforts and items used to counter bacterial infection. Used in the context of waste water treatment, the dictionary lists "septic tank" as "a tank in which the solid matter of continuously flowing sewage is disintegrated by bacteria."

To more specifically describe that process, engineers would use the term "digestion," defined as "the process in sewage treatment by which organic matter in sludge is decomposed by anaerobic bacteria with the release of a burnable mixture of gases." The significance of specifying anaerobic digestion, meaning "in the absence of free oxygen," is not lost on anyone who has been in the presence of aerobic ("in the presence of oxygen") digestion, which is accompanied by the emanation of strong odors. Centralized waste water treatment plants use both aerobic and anaerobic digestion, in different phases of the treatment process.

Septic systems provide a means for dealing with waste water in facilities that are not served by sewage collection and treatment systems. Consisting of a tank, an effluent distribution system and an absorption field, such systems typically are used for outlying or remotely located homes and small service or commercial structures equipped with running water. In addition, septic systems may be appropriate for some public and recreational facilities.

As an extension of the plumbing system, beyond the structure itself, a healthy septic system enables a building's plumbing system to function properly, without backing up or fouling the atmosphere. It does this by providing a buffer for the sewage leaving the building before it enters the surrounding ecosystem. (The unbuffered system in use prior to the development of septic systems was the cesspool, simply a collection spot, such as a hole in the ground, with no provision for long term performance.)

As waste water flows into a septic system, the septic tank provides a settling time for the raw sewage entering the system, after which bacterial action reduces the volume of remaining solids. The bacterial action also begins breaking down other organic components in the waste water, converting potentially harmful materials into more benign remnants. Once the effluent leaves the system through distribution into the absorption field, natural processes continue to break down any remaining organics and the soil filters out remaining particulates as the water migrates toward the local water table.

Role of Plumbing Engineers

The essence of septic system design is balancing inflow and outflow, with a suitable settling and treatment time between when waste water enters the septic system and when it leaves as effluent. Domestic sewage is about 99 percent water. Accordingly, system operation is governed by the performance of the absorption field, which can vary widely because of the highly variable nature of soils and because it is a dynamic system, which is to say it changes over time.

Where additional solids or other materials such as grease or oils are a normal part of the waste water stream, special purpose separators should be provided in the system to facilitate their removal. From an engineering perspective, septic systems cross over into several disciplines, mostly rooted in civil engineering. Plumbing engineers with a limited background or little experience in hydrology, soil mechanics and environmental engineering should proceed with care in working with septic systems.

An appropriate role might be to begin with determining the expected flow into the system and end with coordinating the input of other experts in these fields. Another interesting twist is that septic systems, as well as other waste water treatment systems, tend to fall under the jurisdiction of state or local health departments, rather than building or plumbing departments. This is all the more reason to enlist the help of professionals specializing in this field. Even so, an understanding of the processes and dynamics involved can prove very useful in designing plumbing systems that work well with septic systems.

System Components and Design

Conventions and common practices related to septic systems vary significantly by locale. Influencing factors include weather patterns, such as frost depth and rainfall patterns, as well as soil composition and surface profiles. Once in service, septic system performance depends heavily upon regular maintenance, which includes pumping out the tank. It also is important to prevent the introduction of materials that would be detrimental to the bacteria at work in the system, such as harsh chemicals or antibacterial soaps.

Limiting or eliminating extra solids is also advisable. For example, many sources recommend against using kitchen garbage disposers with septic systems.Septic system capacity is controlled partly by component size considerations and partly by the speed of the natural processes relied upon to break down the waste. Capacity requirements are usually set by the local health department and are typically based on the size, use and occupancy of the facility. When the design flow indicates the need for a septic system too large for the site, other waste water treatment systems must be considered.

The building sanitary drainage system empties into the septic tank. Buried somewhere near the structure, this tank should be watertight and resistant to corrosion and decay. Internal baffles are placed in some tanks to control the flow, channeling incoming waste water into the settling chamber and containing the sludge. Inlet and outlet piping is situated to allow effluent to leave the tank while retaining settled solids (sludge) and scum floating on the liquid surface. Tank sizes range upward from 500 gallons, which is on the small side. Septic tanks traditionally have been constructed of concrete, either precast or poured in place. In recent years, tanks constructed of fiberglass and plastics have come into more widespread use.

The second component in the system is the effluent distribution system. Once the system goes into operation, for every inflow into the septic tank an equal amount of effluent is discharged. For simple systems, the effluent flows directly into a configuration of perforated subterranean piping designed to spread the effluent over a large area (the absorption field). Sometimes soil conditions or the effluent distribution system configuration dictate that a distribution box be employed to split the effluent, discharging it into several field disposal lines.

Another alternative is a dosing chamber, where a pump can be used to send the effluent through the delivery lines into the absorption field. Location and soil suitability are the primary concerns with regard to the absorption field. "Septic Systems: What You Need to Know" says "Generally, at least one acre of land is needed for a properly functioning septic system." One important consideration is the location of the absorption field in relation to wells and other items.

Specific minimum horizontal distances from a septic system to various items. For example, a septic tank should be at least 500 feet from a public water supply, 50 feet from a well or suction line and 15 feet from a dwelling. The absorption field should be a minimum of 500 feet from a public water supply, 100 feet from a well or suction line and 30 feet from a dwelling. Percolation tests and other soil evaluation tools are used to determine the suitability of a particular site, all as dictat - ed by local authorities. Where flow rates or site conditions dictate, alternative types of septic systems can be considered. These include low-pressure pipe systems, fill systems and aerobic treatment units. They typically cost more than conventional septic systems, and require added inspection and maintenance.

Other options for onsite waste water disposal include cluster systems, sand filters, mound systems and spray irrigation systems. Sand filters to be included in the system beyond the septic tank, for the specific purpose of removing nitrogen from the effluent.

Another alternative to a pure septic system for onsite waste water treatment is a mechanized systems. For performance enhancement, some sources suggest additives that stimulate bacterial action.

7

Ecological Genetics

Ecological genetics dealt with the "adjustments and adaptations of wild populations to their natural environment. By growing plants in common gardens at different altitudes, they were able to analyse variation (i.e., local adaptation) that would otherwise have been masked by the use of a single common garden. This was the theme later taken up by gene ecologists. In animals, the link between ecological and genetic variation was thoroughly investigated in the banded snail, Cepaea nemoralis.

The two essential and inseparable aspects of ecological genetics, the combination of ecological fieldwork and laboratory genetics. It was through these types of studies that major syntheses of ecological genetics and evolutionary biology have been made. If ecological genetics is to be successfully developed, we need new thinking, new methods, new material." Over the last 25 years, new thinking, new methods and new material have indeed become available and are being successfully used to develop ecological genetics, much of which has now been encompassed by the phrase "molecular ecology".

The renewed interest in the field can be measured in part by the increased incidence of meetings tackling ecological genetic issues and the success of a journal dedicated to the area, only initiated a decade or so ago. The conceptual developments that have had substantial impacts on the development of ecological genetics have been in areas as diverse as the development of rigorous methods for phylogenetic analysis, the development of procedures for the detailed analysis of gene flow and the development of metapopulation theory.

The greatest impact of new methods has been the development of protein and DNA markers for the analysis of genetic variation. Such laboratory advances have meant that vastly greater numbers of increasingly more variable loci can be identified and used; this shows no sign of slowing. In addition, there have been methodological advances in the statistical analysis of quantitative characters.

Furthermore, low-cost computing power and the concomitant development of analysis software, much of which is web-accessible, have opened up the possibilities of the complex analysis of vast amounts of empirical data. These developments mean that detailed genetic analyses are no longer restricted to an elite group of organisms (e.g., mice, Drosophila, maize); now ecological genetic procedures and analysis should, at least in theory, be open to any organism.

Classical ecological genetics was principally concerned with documenting and measuring the magnitude of selection in natural systems. Such investigations tended to focus on single populations, in which population size was not an essential feature, and on aspects of the phenotype under simple gene control. In contrast, the contemporary approach emphasizes the importance of population size and structure, the interaction among populations through migration and dispersal (i.e., gene flow), the interaction between local selection and genetic drift, and an expansion of the phenotype to include quantitative and qualitative features.

However, into this brave new world, the choice of material for ecological genetic investigations is of the utmost importance, is still valid, albeit with modifications in the light of nearly 30 years of additional research and development. Thus, it is necessary to have a clear, testable hypothesis. In addition, the ideal study taxon should:

— have a known ecology;

— be easy to collect;

— have a short life-span so that numerous generations can be investigated;

— have genetic variation that is easily recognized and interpreted and is capable of statistical analysis;

— be at a high enough density that it can easily be found; and

— be easily maintained in the laboratory or greenhouse.

However, it is rarely possible to fulfill all of these criteria. The choice of material seems "rather an art than a science", and stated that "the more rewarding decisions are generally reached by a thorough grasp of the essentials of the problem to be solved, estimated in the light of much knowledge and experience".

Lifetimes of research are often required for robust investigations of ecological genetics, and require a combination of detailed fieldwork and the application of genetic (now usually molecular) techniques. However, importantly ecological genetics requires experimentation. Documentation of variation needs to be combined with the generation of hypotheses that can be empirically tested.

We have interpreted contemporary ecological genetics as the investigation of the origin and maintenance of genetic variation within and between populations, which ultimately leads to adaptation and speciation. Ecological genetics provides the means by which mechanisms of speciation can be understood and the patterns of genetic variation in nature investigated.

Modern Ecological Genetics

The picture of modern ecological genetics that emerges is one of relevance to the interaction of large-scale geographic patterns of demography with genetic dynamics among small, partially isolated, and potentially locally adapted populations. Framed in such a context, ecological genetics is of both great pure and applied scientific interest.

Adaptive radiation, the evolution of ecological diversity within a rapidly multiplying phylogenetic lineage, is one major research area where the combination of ecological and genetic analysis has been particular valuable. The debate over the introduction of alien species and release of genetically modified (GM) organisms and their potential impacts on the environment and closely related species is an area where a combination of fieldwork and experimental work, within an ecological genetic framework, is likely to produce exciting pure and applied science.

One of the major environmental concerns over GM crops relates to transgene movement into wild relatives, where the pattern of hybridization will affect the scale and rapidity of any ecological change, plus the feasibility of any necessary containment. In the UK, one of the species of most concern is the widely cultivated oilseed rape (Brassica napus), since it has numerous wild relatives (e.g., Brassica rapa) that are either native or naturalized. Using remote sensing, Wilkinson et al. identified possible sites of sympatry between B. napus and B. rapa across 15,000 km^2 of south-east England.

In 1998, two sympatric populations were found over the entire survey area, and in 1999, every newly recruited plant in these populations was tested for hybridity using a combination of flow cytometry and nuclear DNA markers. Despite the size of this investigation, only one hybrid was observed among 505 plants screened in the B. rapa populations. Thus, it is possible to investigate the interaction between populations of domesticated and wild organisms. As human impacts on habitat landscape increase, and habitats are fragmented, the interactions between wild and domesticated populations become more complex.

Similarly, wild populations that were once large and connected by gene flow may become fragmented, and the population size diminished. Under such conditions, migration becomes critical for the reestablishment of local populations, so that gene flow may counter some of the detrimental effects of inbreeding depression. Similarly, changes in habitat management may have significant influences on the genetics of rare species. For example, in the UK, the Red Squirrel (Scirius vulgaris) is a nationally rare species, although it is locally common in the fragmented coniferous woodlands of northern England and the Scottish Borders.

The creation of a large coniferous plantation was enough to produce miscegenation of S. vulgaris populations up to 100 km apart. Thus, the genetic consequences of habitat change may occur hundreds of kilometers away from the site of change. Any ecological genetics program that focuses on demographic and genetic variation across large-scale geographic variation has obvious implications for conservation. Populations that are small and locally adapted may also suffer from increased gene flow through the disruption of locally adapted genotypes. However, any migrants into these populations may not have equal effects on local population growth and fitness.

Thus, knowledge of the genetic diversity that exists among migrants is essential when constructing conservation policies for species reintroduction or augmenting declining populations. For example, conservation organizations are increasingly interested in ensuring that only local seed is planted within specific areas. This, of course, raises the issue of what is meant by local: the same population, the same county, the same country or even the same continent.

Conservation biology and GM risk assessment are not the only areas where the results from ecological genetic investigations are a valuable applied tool. The long-term sustainability of ecological systems depends on the ability of organisms to adapt and respond to changes in their biotic and abiotic environment.

The methods for studying adaptive evolution in contemporary populations provided by ecological genetics are equally important in assessing the ability of organisms to respond to both local and global environmental perturbations. Furthermore, it is now possible to investigate the genetic and ecological outcomes of major environmental changes, for example, the effects of the last glaciation on the recolonization of Europe by plant and animal species. The accurate reconstruction of historical species' range shifts and correlation with past climate change offer great potential for improving predictions of floral and faunal adaptive response and gene flow under future rapid climate change scenarios.

New developments in DNA extraction technology have been incorporated in ecological genetic approaches. It is no longer necessary to collect whole organisms; all that is required is a small amount of suitable material. Furthermore, the use of preserved materials means that it is possible to add a temporal dimension (tens, even thousands of years) to investigate genetic diversity and structure. Also, hundreds of samples may be readily analyzed for large genomic segments. Thousands of markers are available, which are variable at many different levels.

However, the next major development will be through environmental genomics initiatives, where the expression patterns of genetic variants in different environments can be investigated. Genetics is here to stay as an essential component of ecology, while ecology will become increasingly important for the interpretation of data emerging from genomics research programs.

8

Genetically Modified Micro-organisms and Viruses

Genetically modified micro-organisms have been used for 20 years in the production of pharmaceuticals (eg. human insulin, bovine somatotrophin), vaccines (eg. the hepatitis B vaccine) and enzymes (eg. rennet for the production of cheese). However, none of these applications involve the release of genetically modified micro-organisms or viruses. Applications that involve the release of genetically modified microorganisms or viruses into the environment will be considered.

Micro-organisms and viruses poses special risk-assessment problems:

— Monitoring the spread of the micro-organisms or viruses is difficult.

— The effect of the micro-organisms or viruses on nontarget species can be difficult to evaluate.

— Large numbers of micro-organisms have to be released for many applications.

— Some micro-organisms can multiply rapidly.

— Horizontal gene transfer is widespread in bacteria. If the genes that are introduced into the micro-organism are on mobile genetic elements such as plasmids, transposable elements or bacteriophage, there is a risk of transfer to other bacterial species. The risk of horizontal gene transfer can be addressed through the design of safer systems for the construction of genetically modified microorganisms.

Some of these risks are not peculiar to the release of a genetically modified micro-organism or virus. Some of the potential applications of gene technology that involve the release of genetically modified microorganisms or viruses are listed below. In a small number of cases, release of genetically modified micro-organisms has already received approval, in Australia or overseas.

Release of Genetically Modified Bacterium

Frost Injury

The US Environmental Protection Agency granted the first approval for the release of a genetically modified bacterium in 1985. The bacterium was Pseudomonas syringae which had been altered to prevent the production of a protein that promotes the formation of ice crystals on plants. The genetically modified bacteria did not contain any foreign DNA and after 18 days could not be detected in the soil.

The bacterium is normally found in the environment. A potential risk might be the loss of normally occurring frost injury in non-target plants, especially if the altered bacterium could spread. If frost is an important limiting factor in the propagation of non-target plants, the spread of ice-minus bacteria could have an environmental impact.

Plant Diseases

Agrobacterium radiobacter is used in the biological control of crown gall, a bacterial disease of stone fruits, nut trees and roses. Unmodified A. radiobacter produces a natural antibiotic that is crucial to its biocontrol activity. The genetic modification prevents the transfer of genes involved in the production of the antibiotic to other soil bacteria and thus prevents immunity to the biocontrol agent. The modified bacterium does not contain any foreign DNA.

Other potential biocontrol agents are under development (eg. modification of Pseudomonas species for the control of fungal diseases. The use of these genetically modified micro-organisms has the potential to reduce use of chemicals for the control of bacterial and fungal diseases. If the engineered bacteria do not persist in the environment, the effect on non-target species would be no greater than when chemical methods are used. However, if the genetically modified microorganisms do persist, the environmental consequences would be more difficult to predict.

Insect Pests

A number of different strategies involving use of genetically modified bacteria and viruses to control insect pests are under investigation. Some of these, such as Bacillus thuringiensis (Bt), involve expression of microbial genes by GMO plants. Another area of research interest is the genetic modification of baculoviruses - viruses that infect insects. Though baculovirus infection is usually lethal, their use as biocontrol agents has been limited due to the length of time it takes to kill the insect.

Genes that encode various insect toxins (eg. Bt toxins) could be inserted into baculoviruses to make them more effective as biocontrol agents. As for the control of plant diseases, the use of these genetically modified micro-organisms and viruses has the potential to reduce use of insecticides, with substantial environmental benefits. If the engineered microorganisms and viruses do not persist in the environment or are specific to an introduced pest species, the effect on non-target species would be no greater than when chemical methods are used. However, if the genetically modified micro-organisms do persist and/or are nonspecific the environmental consequences would be more difficult to predict.

Plant-associated Microbes

Bacteria that have a beneficial effect on plant growth (eg. nitrogen-fixing bacteria and bacteria that produce plant hormones) are potential targets for genetic modification. In 1997, the US Environmental Protection Agency approved the commercial release of a genetically modified Rhizobium melitoli, which contains extra copies of two genes that enhance nitrogen fixation. The producers of Dormal PLUS™ claim that lucerne seed inoculated with Dormal PLUS™ has a yield increase of 6% over that inoculated with unmodified Rhizobium.

The genes involved in colonisation of plant roots, with the aim of engineering nitrogen-fixing bacteria which can colonise non-leguminous plants. Increased productivity of crop plants could potentially reduce land-clearing. However, if modified bacteria were able to colonise non-target plants there would be serious concerns about the possible environmental impact. The use of genetically engineered 'immunocontraceptive' viral vaccines for the control or mice, rabbits and foxes. The control of introduced pest species has obvious benefits for the environment. For such a strategy to be accepted, the virus must be absolutely specific to the host target.

Rumen Micro-organisms

In ruminants, a high proportion of energy intake is lost due to inefficient digestion of cellulose. Modification of rumen bacteria to improve digestion of cellulose is one focus of research. The strategies include introducing genes for cellulolytic enzymes which function at low pH and genes which allow rumen bacteria to degrade xylose and pectins.

The use of genetically modified rumen micro-organisms could have an environmental impact if the increased feed efficiency of ruminants led to increased stocking rates and greater use of marginal lands. If the modified bacteria could colonise native animals, it might impact on their feed efficiency and could, as a consequence, lead to enhanced reproductive capabilities. However, the trials to date suggest that genetically modified micro-organisms compete so poorly in the rumen that their potential usefulness is questionable.

Vaccines

Genetically modified viruses and bacteria are being developed for the prevention of human and animal disease. Some examples include the development of a recombinant vaccinia virus-rabies vaccine and a recombinant porcine adenovirus swine fever virus vaccine. The safety of such vaccines would have to be thoroughly evaluated before release, especially if vaccinated individuals will shed infectious virus or bacteria.

Bioremediation

The use of micro-organisms for chemical decontamination has been called bioremediation. Many micro-organisms are naturally able to transform toxic chemicals into harmless products. The aim of genetic engineering research is to increase the efficiency of the chemical transformation process or increase the range of toxic chemicals which are degraded. For example, the US Natural and Accelerated Bioremediation Research (NABIR) program is aimed at the development of bioremediation techniques for the removal of radionuclides and metals from the 7000 US Department of Energy sites where subsurface contamination, generated during 50 years of research, development, and testing of nuclear materials, has been detected.

Soil Microbes

The specific roles of soil GMOs for mediating processes that enhance crop yields and for degrading hazardous environmental pollutants are reviewed, along with the valuable role that this technology can play in increasing our understanding of the structure, processes and ecology of soil microbial communities. Many of the methods developed for detecting GMOs are now being used to monitor the survival of soil microbes, and to identify their niches and roles in the complex community of the soil food-web.

The fate of microbial inocula introduced into the soil environment is less predictable than the introduction of plants. For microbial introductions to prevail, they must be able to compete successfully with highly diverse indigenous populations that are well adapted to soil type, and to survive under stress and subject to a wide range of natural predatory soil fauna. Microorganisms that function within a symbiotic relationship with plants would appear to be better subjects for engineering and introduction than free-living, self supporting GMOs.

One early example of symbiont success was an engineered strain of Rhizobium meliloti that improved nitrogen fixation and growth of lucerne in field tests. The eukaryotic fungi differ both anatomically and physiologically from the prokaryotic bacteria. Nevertheless, as heterotrophs they can effectively combine in functionally important soil processes such as the decomposition of organic detritus. Interspecific competition within each phylum is strong.

Nonengineered fungal species have been used as effective control agents of soil-borne plant diseases since the 1970s. A cross-phylum example has been the antagonistic colonisation of the take-all fungus in wheat by rodshaped bacteria in rhizosphere soil. The engineering of existing species of both soil fungi and bacteria might be able to promote disease control, but there is a general view that the prokaryotes present better opportunities for genetic manipulation.

GMOs can enter the soil in a variety of ways, including wash-down of microbes applied as foliar treatments for crops. Another external source could be engineered chemo-autotrophs (eg. Thiobacillus), which are used in the mining industry to accelerate the leaching of metal from ores. However, specialised microbes survive in the competitive soil environment and it is also unlikely that the planned introduction of GMOs will have a significant effect on non-target wild strains and their essential servicing functions.

Only one or a few inserted genes among thousands would be involved and, in any case, gene interchange is a common phenomenon in the soil microbial community. However, because of their novelty, there is need to evaluate risks associated with their release into the environment and to assess their survival and effectiveness in expressing their designed function.

Soil Bioremediation

Within the well-adapted soil community there is substantial transfer of genes, by conjugation (transfer of bacterial plasmids - chromosomal free DNA), transduction (bacteriophage mediated) and transformation (uptake and expression of extra-cellular DNA). This lability presents a problem for the introduction of soil GMOs, as the dispersal of their genes will also be fostered by interactions with other soil biota as well, and by the physico-chemical environment eg. water movement. Detection and monitoring of GMOs require a wide range of techniques for detecting single cells in situ, assaying their activity and monitoring their dispersion.

A marker system based on insertion of the lux gene (from marine organisms) into selected soil inocula, has been effective, because soil provides little background luminescence. The lux gene and other insertions are replacing the earlier and publicly sensitive use of antibiotic resistance markers. Hot spots of antibiotic resistance genes have been identified in all soil environments. It has been recommended that the use of genes encoding resistance to all clinically relevant antibiotics should be avoided.

The fact that only a small proportion of bacteria is readily accessible by standard culturing techniques, and bacterial cells lose their ability to grow on solid media after exposure to environmental stress, severely complicates assessment of their environmental fate. However, there has been rapid progress in developing new tools and techniques to fill the void. Phylogenetic identification and classification of bacterial and fungal hosts is critical for safety evaluation.

The use of 16S rDNA fragments, amplified from directly extracted nucleic acid genes as a molecular marker, is now an established and successful method for determining phylogenetic relationships, and sensitive and highly effective software for assigning sequences to a 'genetic tree' is a major achievement. Also, combining advanced microscopic technique with either reporter genes or fluorescing probes, has facilitated in situ analysis of microbes.

Plant GMOs

Random genetic changes occur spontaneously in wild plant species. People have been engaged for centuries in plant breeding using selective crossing to improve a beneficial trait, followed by back-crossing to a parent to eliminate unwanted genes. Genetic engineering can target desirable traits more precisely, by transferring specified genetic material into plant cells. By exercising this greater control over the process, improved plant performance can be achieved more rapidly.

The success rate for engineered plants is continually being improved by growing access to genomic libraries and by newer techniques for transferring genes into unrelated species to express novel traits not enabled by traditional breeding techniques. Laboratory-based transfer may involve cellular uptake of the new DNA using a carrier or by the fusion of whole cells. Tissue culture is then used to encourage the growth of modified cells to produce a fully functional plant. GMO success depends on the choice of genes for transfer, their expression and on the interaction between inserted genes and the cellular environment in which they reside.

Wheat has been the staple grain crop in Australia since the 1860s, with management and genetics being the drivers of yield changes. Between 1860 to 1900, yield declined from 1 to 0.5 tonnes per ha and this can be attributed to nutrient exhaustion and to mismatch between imported varieties and the Australian climate. Over the next 50 years, with combined use of superphosphate, fallowing, new varieties, and movement of cropping to the slopes and plains, yield recovered to a level of 1 tonne per ha. Between 1950 and 1980, legume nitrogen and rotations, mechanisation, and the introduction of semi-dwarf strains increased production to 1.4 tonnes per ha.

By 2000, yield had increased to 1.8 tonnes per ha and this was substantially due to the increasing use of leguminous grain crops in the rotation. Such enhanced cropping diversity in the farming enterprise can provide improved soil fertility and financial stability. It also provides a greater market for new GM cultivars, and it is likely that further diversification will be provided by subsidiary GM varieties, wherein the chemical nature of the food provided will produce a range of health-enhancing products. Examples are improved nutritional balance, polyunsaturated fats, reduced allergens and vitamin supplementation.

The Plant Biotechnology Centre has a novel project to enable 'on-off ' switching of the lignin gene in pasture grasses. Success could either increase the 'toughness' of grasses under recreational use or increase the 'softness' of grasses by reducing lignin content, which contributes to a low nutritive value for ruminants. The Plant Molecular Biology and Technology Unit has developed a hypoallergenic rye-grass, and has a project to develop a vaccine for control of grass pollen induced hay fever. It also has reported a successful tissue culture system for propagating macadamia - the sole Australian native species to be commercialised as a crop.

Pre-product planning:

— Assess status of intellectual property (IP) and 'freedom-to-operate' (Patents).

— Preliminary tests for potential allergens, toxicities and other anti-nutritional effects including undesirable protein modifications.

— Devise a strategy for a full evaluation of food safety.

First-stage product development

— Optimise gene constructs.

— Modify regulatory systems and coding sequences to produce desired expression level and its temporal and spatial expression pattern.

— Identify best transgenic lines based on sufficient events to base evaluation.

— Insertion events, followed by stability evaluation over a minimum of three generations.

Field testing

— Test the agronomic performance of chosen lines in multiple field trials.

— Segregation-based removal of the first-stage selectable markers.

— Development of an effective marker for following the transgene in breeding material.

— Multiple field trials to determine the quality of seed from the GM plant.

Release to Evironment

The Ecological Society of America (ESA) is a prestigious body representing 8000 ecological scientists. In a recent policy statement, it supports the judicious use of biotechnology. The following are extracts from this statement.

GMOs have the potential to play a role in sustainable agriculture, forestry, aquaculture and bioremediation. However, both deliberate and inadvertent releases of GMOs into the environment could have negative ecological impacts under certain circumstances. Engineered organisms that may pose some risk and hence require scrutiny include cases where there is uncertainty about environmental effects. These could be cases where:

— There is little prior experience with the organismal trait and host combination

— An organism may persist without human intervention

— A genetic exchange is possible between a transformed organism and other organisms

— The trait confers an advantage to the GMO over native species in a given environment An assessment of environmental risk is needed to minimize the likelihood of negative ecological effects such as:

— Creating new and more vigorous pests and pathogens Exacerbating the effects of existing pests through hybridization with related transgenic plants or animals

— Harm to non-target species, such as soil organisms, non-pest insects, birds and other animals; and Irreparable loss or change in species diversity and genetic diversity within species.

Given that losses in diversity needed to be clearly associated with introduction of GMOs and not other aspects of management, then these categories present a reasoned assessment of risks to the environment, which need scrutiny in any determination of net benefit. The following protocol, provides an example of the controls applied in Australia for minimising the risk of gene escape from GM plants. A range of controls can include the following:

The inclusion of buffer crops to act as pollen traps: Some species have a low tendency to disperse and only small buffer zones may be required. Crops with lighter weight pollen, easily wind dispersed, may need either larger buffer zones, or pollen containment by placing polythene bags over the flowers before maturity. Crop management practices, such as pruning can minimise flower formation. Where flowering and seed production are not required, plants can be harvested before flowering so that seed and pollen are not produced.

Animal and bird proofing and/or insect minimisation: Seeds can be harvested by hand to avoid seed dispersion. Postharvest clean-up procedures include the incorporation of plant residues into the soil, followed by at least a two-year monitoring period to check whether the environment may have been contaminated. Buffer plants are usually destroyed after the trial. At times, they are examined first to determine the extent of 'gene flow' from the trial plants to the buffer plants.

Deleterious gene escape from GM plants has already occurred. A recent international case, occurred this year in Canada and involved escape through pollen movement. Cross-pollination from GM canola varieties led to triple herbicide resistance among canola varieties, which should serve as a warning to producers to use their new herbicideresistant varieties wisely according to the guidelines. In Australia, GM canola (produced by Monsanto and Aventis) trials were conducted in Tasmania from 1996-99 and for a further three years to evaluate environmental persistence.

The trials from 1996-99 reached seedproduction scale, with a voluntary protocol provided by GMAC. This included advice to minimise the risk of trans-genes and gene flow from the trials to other crops and to weeds of the Brassica genus. In the post-trial threeyear period, the then interim office of the gene technology regulator (IOGTR) identified non-compliance, and ordered remedial action and an extension to the trial period.

Non-compliance was found at 21 of the 57 sites. Each of these sites was deemed noncompliant based on the discovery of one or more canola volunteer plants that had flowered or set seed or on which the seed pods had shattered on or close to a site within three years of the crop.

Insect-resistant GM plants

Breeding plants which resist insects is not new. Before World War II, commonly used crop varieties had varying levels of resistance to the pests they had co-evolved with. Exploitation of this resistance, either deliberately or unconsciously, was a major component of insect pest management strategies of the time. Some successes had been achieved in breeding for pest resistance by conventional means.

After World War II, with the widespread adoption of cheap and effective synthetic organic insecticides such as DDT and the organophosphates, breeding resistant varieties received less attention. Indeed, existing sources of resistance were lost in some cases. The case of gossypol (and related terpenoids) in cotton is instructive. These secondary compounds provide resistance to most chewing insects. Wild Gossypium species, which contain them in abundance, are little affected by the pests of cultivated cotton.

However, gossypol is toxic to monogastric animals, and if cottonseed meal is used for pigs and poultry it must be heated, leading to a loss of protein quality. Gossypol was therefore deliberately bred out of American cotton varieties in the 1950s, and these varieties went on to form the basis of elite cotton lines in many countries around the world. High gossypol lines now have poor yield, poor quality and undesirable agronomic characteristics compared with current elite varieties.

Even if insect-resistance traits are not deliberately selected against, as in the case of gossypol, they can be accidentally lost during the course of breeding if yield and quality are the primary selection criteria. Variety trials have commonly been conducted under intensive insecticide coverage, and in these circumstances the most favoured genotypes are likely to be those which devote more of their resources to yield, perhaps at the expense of secondary chemicals which provide resistance to pests.

This may explain why in soybeans, for example, insectresistant varieties have 10-40% lower yields than susceptible ones, in the absence of pests. There has been no deliberate selection against insect resistance in soybeans, in contrast to the gossypol case in cotton. The legacy of this historical trend of accidental or deliberate loss of resistance is that, within plant species, sources of resistance tend to be in older and poorer varieties. This makes breeding for insect resistance a tedious and costly process.

The process has nevertheless resulted in some significant successes for Australian agriculture in the last three or four decades. Major pests such as the sorghum midge, Contarinia sorghicola, and the spotted alfalfa aphid, Therioaphis trifolii maculata, have been brought under control by breeding or importing resistant varieties. Integrated pest management programs which have a resistant variety as a major component are usually stable and strong.

Role of Biotechnology

Breeding for insect resistance using conventional methods of repeated backcrossing and selection will continue to have a role in insect pest management for Australian agriculture. In cotton, for example, there are many morphological and biochemical traits which confer resistance to many significant pests, and considerable variation among current varieties in resistance to these pests. Resistance incorporated by conventional breeding is frequently polygenic, and therefore less susceptible to resistance-breaking pest biotypes.

Nevertheless, biotechnology, in the form of genetic manipulation, has special advantages for developing pestresistant varieties. It offers a quick way of incorporating resistance into current elite varieties, avoiding tedious backcrossing and selection. Given the situation outlined previously, leading to the concentration of resistance traits in poor varieties, this is a considerable advantage.

The downside of this is that the traits most suitable for incorporation using current biotechnology are simple monogenic ones. This leaves them vulnerable to resistance-breaking biotypes, unless they are carefully managed. The other major advantage of biotechnology over conventional breeding is that the store of resistance traits is not limited to the gene pool of the plant species. Genes from other higher plants, or from other organisms entirely, can be exploited. This greatly expands the potential for breeding resistant varieties.

Bacillus Thuringiensis (Bt)

Bacillus thuringiensis (Bt) is a common spore-forming soil bacterium that produces a crystalline protein toxin (known as a δ endotoxin) when it sporulates. There are many strains of Bt and many variants of the Bt toxin. A genetically modified strain that produces several different toxins has been created. The toxins in this strain are also more resistant to ultra-violet light. Because the strain does not produce viable spores it cannot compete with naturally occurring Bt.

Other researchers have investigated increasing the production of toxins in Bt to reduce production costs and transferring Bt toxin genes to other bacteria (eg. to soil bacteria to provide control of insects that attack plant roots). Bt toxins are grouped into families designated Cry1, Cry2 etc. There are currently over 130 Bt toxins known, and this total is probably conservative because many recent discoveries are commercial-in-confidence. Each family tends to be specific to a given order of insects. For example, Cry1 toxins have most activity on Lepidoptera.

This is because they bind to specific sites on the gut epithelium, leading to rupture of the cells and invasion of the insect haemocoel by secondary pathogens. This specificity is valuable in insect pest management, because it leaves beneficial insects (predators and pathogens) relatively unaffected. Bt has long been used in organic and conventional agriculture, in the form of foliar biopesticides containing Cry1 toxins and spores of Bacillus thuringiensis var. kurstaki. There were 182 Bt products registered by the United States EPA in 1955, though they never constituted more than a few percent of the total sales of insecticides. In Australia, some of these formulations have been widely used against Helicoverpa spp. on cotton.

In this role they have been limited by their slower killing speed compared with many conventional pesticides, but their advantage has been 'softness' for natural enemies. Current cotton use is limited by restrictions imposed to avoid the development of resistance to Bt cotton. Other uses of foliar Bt have included application to brassicaceous vegetables against cabbage moth, Plutella xylostella, and cabbage white butterfly, Pieris rapae. It is also widely used in home gardens against many lepidopteran pests.

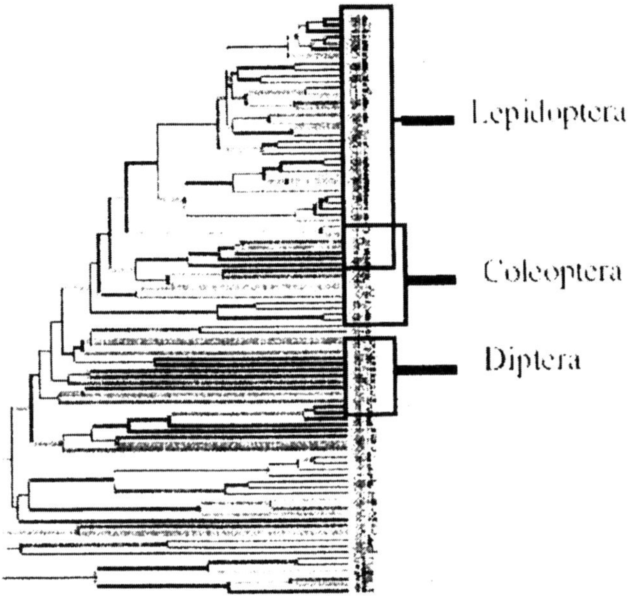

Figure 1. Taxonomy of Bt toxins

Commercial biotechnology became involved with Bt in 1990 when Monsanto patented a gene encoding for a protein almost identical to Cry1Ac, and subsequently introduced it into US cotton varieties under the trade name Bollgard®.

Other Bt genes encoding proteins similar to Cry 2Aa, Cry2Ab (for Helicoverpa spp. in cotton), Cry 1Ab and Cry9C (for European corn borer and other lepidopteran pests in corn) and Cry 3A (for Colorado beetles in potato) have since been exploited. In 2000, Bt varieties of maize constituted 19% of the total acreage in the USA. Bt cotton comprised 39% of the total acreage, ranging from 67% in Mississippi to 3% in California, primarily reflecting the relative importance of lepidopteran pests in these regions.

Adoption of Bt potatoes has been low, at only about 3% of the crop. In Australia, Bt cotton is the only pest-protected GM crop that has been released commercially. It comprises 30% of the total acreage, at which level it is currently capped for the purposes of resistance management.

Other Insecticidal Genes

Most insect-protected crops released around the world have used genes from Bt, and there are still many genes from this source which remain to be exploited (Fig. 1). The specificity of Bt toxins, however, means that there is a limited repertoire for each pest. For Helicoverpa armigera, for example, only four toxins (Cry1Ac, Cry1Ab, Cry2Aa and Cry2Ab) are effective.

However, there is another class of insecticidal proteins in Bacillus thuringiensis, the Vips (vegetative insecticidal proteins), which are found in vegetative cells of B. thuringiensis (and other Bacillus species), as opposed to spores. It is likely that these will prove to be a diverse and potentially useful category of toxins for genetic manipulation, but their effects on non-target organisms remain to be clarified.

There are several alternative sources of resistance, mostly derived from plant genes. Lectins, especially those expressed by the GNA (Galanthus nivalis agglutinin) gene from snowdrops, have long been considered as potential candidates for genetically engineered plants. There are, however, no commercial plant varieties with this GM. Lectins have undesirable effects on mammals and other non-target organisms, and are unlikely to obtain approval from OGTR in Australia. Proteinase inhibitors, especially the trypsin inhibitor from cowpeas, are also long-standing candidates for genetic manipulation.

Many plants have been transformed with the cowpea trypsin inhibitor gene, but there are few commercial releases (and none in Australia). A small percentage of the large area of Bt cotton in China is made up of varieties which also express cowpea trypsin inhibitor. The problem with proteinases has been that the target insects readily adapt to them through increased production of the protein digesting enzymes. This is a form of induced tolerance rather than genetically selected resistance. However, potential remains for exploitation of these genes.

In Australia, Hexima Ltd, a joint venture between the University of Melbourne and Pivot Ltd, an Australian agricultural company, has patented a gene for a proteinase inhibitor derived from wild tobacco, and is working on modifications to its structure for potential use against insect pests in cotton and other crops. Other genes that target selective features of insect metabolism include chitinases, which affect chitin, the primary component of the insect exoskeleton material, as well as cholesterol oxidases and alpha-amylase inhibitors.

Further plant defence chemicals for which the genes have been identified and which could be incorporated in transgenic varieties include the drimanes, terpenoids which have antifeedant actions, especially on aphids, and the cyclic hydroxamic acid, DIMBOA, which confers resistance to the European corn borer in maize. The H. armigera stunt virus is a very small RNA virus particle specific to H. armigera, which can be incorporated in the genome of cotton, and may be useful in future GM varieties. However, current varieties with this transformation have not given sufficiently high expression to be useful.

The only insect-resistant transgenic crop released commercially in Australia, Bt cotton, would appear to be an almost unqualified success. For example, in the central and northern regions of NSW, endosulfan contamination in waterways has declined markedly in the same time frame as Bt cotton has been taken up. Not all of this has been due to Bt cotton - restrictions on certain methods of application of the chemical, the development of substitute chemicals, and adoption of IPM by the cotton industry have all helped. Nevertheless, Bt cotton has had significant direct and indirect influences on this trend.

This role might extend to the commissioning of research in the area. While there is a plethora of organisations conducting research on water quality, it often seems not to be quantitatively related to changes in agricultural practices, including GM crop adoption. On the other hand, other Bt or pest-protection GMs might not be as benign as Bt cotton appears to have been. An example is the proposal for Bt eucalypts.

An example is the current trials of Bt cotton in northern Australia. Previous attempts to grow cotton in northern Australia were abandoned after the disastrous pest crisis in the Ord River area in the early 1970s. However, new systems are likely to enable cotton growing in these regions. Bt cotton is likely to be a key factor in these systems. Cotton might be produced with fewer environmental problems relating to pesticides than current horticultural crops (such as melons) in these regions. However, this will mean that cotton will be grown where it was not grown before and this will have many potential environmental impacts relating to land and water quality.

Weeds and GM Crops

Transgenic modifications have the potential to radically alter crop-protection strategies used in Australian agriculture where pesticide sales alone cost farmers $1550m annually. Emphasis has been given to 'pest protected' crop varieties, such as Bt cotton, and to 'herbicide tolerant' crop varieties. While the former aim to reduce reliance on pesticides, herbicide tolerance is designed to enable the more widespread use of certain chemicals for weed control.

However, transformation studies have also focused on the herbicide-resistance trait because the physiological basis is well characterised, resistance is often a dominant single-gene trait and it can be used as a selectable marker for transformed plants. Transgenic herbicide tolerant (THT) crops are perceived to offer flexibility in weed management.

Herbicides that lower mammalian toxicity, less residual activity, or have different modes of action to commonly used chemicals (which may help delay herbicide resistance in weeds), may be able to be used 'in crop' that would otherwise not have been possible.

Similarly, THT crops may provide the opportunity to control recalcitrant weeds, such as nutgrass in cotton, as well as a broader spectrum of species. For these reasons, most attention so far has been given to developing transgenic crop varieties which are resistant to the two 'nonselective' chemicals - glyphosate (38%) and glufosinate ammonium (37%). However, for pasture legumes, emphasis has been placed on increasing varietal tolerance to currently used selective herbicides rather than to nonselective chemicals which, if used, could lead to a reduction in species diversity within a pasture system.

Some 90% of varieties of genetically modified crops currently in commercial use throughout the world confer either insecticidal properties on plants (eg. Bt corn and Bt cotton) and/or herbicide tolerance properties and we are now rapidly approaching the situation where it will technically be possible to insert herbicide-resistance genes into all major Australian crop species. In Australia to date, cotton resistant to the herbicide glyphosate, and carnations that contain a marker gene conferring resistance to the sulfonylurea herbicides, are the only GM herbicide tolerant crops to have been released commercially, though several other crops such as sweet lupins modified for resistance to glufosinate ammonium have been approved for experimental releases.

Large-scale field trials have been under way on herbicide tolerant canola for several years and these experimental crops have attracted considerable attention from lobby groups on either side of the GM debate. While the technology for gene transfer for pasture plants is behind that for some of the more important crop plants, rapid progress is being made, such that lucerne (Medicago sativa), subterranean clover (Trifolium subterraneum), white clover (T. repens) and several other species can all now be genetically modified.

Approval has also been given for experimental release of glufosinate-resistant subterranean clover, but Radcliffe believes that herbicide-tolerant clover and lupins are unlikely to be used commercially, presumably because of either market or environmental factors. Based on proposals currently before gene technology regulators in Australia, and experience with herbicide tolerant crops in Canada and the USA, it is likely that there will be many more applications for trials and commercial release of such varieties in Australia over the next 5 to 10 years.

Given the current research emphasis and community expectation for lower pesticide use in agriculture, concerns have been raised over the commercial use of THT crops and their effects on farming systems and the land and water resource base. One concern is that the use of THT crops will lead to increased use of chemicals for weed control with consequent health and environmental effects. This will partly depend on how a herbicide associated with a THT crop substitutes for currently used herbicides.

There are indications from Australia and overseas that the introduction of GM herbicide-tolerant crops may lead to either a decrease or increase in herbicide use depending on the system concerned. The type of herbicide is also important. For example, it could be argued that there would be less environmental contamination with GM glyphosate tolerant canola than with the non-GM triazine tolerant (TT) and imidazolinone ('imi') tolerant canola varieties currently grown commercially in Australia, because the latter varieties rely on herbicides with greater soil residual activity.

Whether or not the THT crops would lead to greater herbicide use if introduced may depend on whether or not they remove a limitation on the area of that crop sown. For example, the introduction of TT canola in 1994 with the ability to control certain weeds was an important contributing factor to the tenfold increase in area sown to canola in subsequent years and the accompanying increase in triazine herbicide use across Australia.

A second concern with THT crops is a shift in the weed flora, for example, to weeds with protracted and late emergence periods, and particularly the more rapid evolution of weed populations that are resistant to the particular herbicide to which the crop or pasture variety is resistant. If THT crop varieties are to be released in Australia and be an effective tool in weed management then they need to be incorporated into an integrated weed management framework where all appropriate weed control options are utilised so that resistance to any single technique does not develop.

If used alone as a single strategy, THT crops and their associated herbicides will fail to be effective in the long term, just as single chemical treatments are currently failing due to the development of herbicide-resistant weeds. Other selective herbicides would remain effective within those crops, albeit against a different spectrum of weeds, though conceivably, if one or two GM-related products swamp the market, competitive products and companies may not survive and then alternative herbicides may not be available when they are ultimately needed.

Another concern is that THT crops may themselves become major weeds, particularly as volunteers in succeeding crop rotations. This is more likely to be a problem where 'volunteer' crop weeds from the preceding crop are normally controlled by the herbicide to which the THT crop plant is resistant. Moreover, recent experience with GM canola trials in Australia has shown that canola may continue to emerge as a volunteer plant for 3 to 5 years after it was first sown. So the persistence and movement of volunteer crop weeds appears to be a valid problem.

One of the greatest risks from the introduction of THT crops into Australian cropping systems is the potential for transfer of herbicide resistance from transgenic crop varieties to their weedy relatives, whether they be related weedy species or weedy races of the crops themselves. For herbicide-tolerance transgenes, the main consequence of such introgression would be the loss of a previously effective chemical weed control strategy. In most cases, this will be to the non-selective 'safety net' herbicides glyphosate and glufosinate.

Introgression will depend on both the presence of plants that could receive the transformed gene and the frequency of outcrossing and sexual compatibility of the species, and yet we have very little data on these aspects to carry out adequate risk assessments for most crops and weeds in Australia. Particular concern has been raised over cross-pollination between herbicidetolerant canola plants and their weedy relatives.

It must be remembered that herbicide resistance in a weed population in non-GM systems usually begins from a very small base level and can increase over a relatively short time to involve a majority of the population when the population is subjected to intensive herbicidal selection pressure. Therefore, it can be argued that THT crops that provide even a slight risk of flow of herbicide resistant genes into weedy species should not be considered.

Intra and interspecific diversity in weed floras is a major constraint to weed management, and the utility of existing herbicides must not be compromised in any way in order that their viability and current spectral efficacy is ensured. The transfer of herbicide resistance to most non-weedy wild or native plants is likely to pose less of a threat for weed management, although there is some concern over genetic contamination in the natural environment.

Herbicide-resistant plants are unlikely to have any competitive advantage in natural habitats where herbicides are not used. It is not only the THT crops that may influence weed management issues. It is conceivable that a scenario may arise where a minor weed which is currently being suppressed by native insects and diseases is no longer subject to such biological constraints because of the introduction of a pest protected GM crop variety and is able then to develop into a major weed pest.

Equally, drought, frost, acid soil, temperature or salt tolerance could be risky traits if they were passed from transgenic crops to their weedy relatives. Weeds may then be more tolerant of a wider range of environmental conditions and may spread to and pose problems in previously uninfested areas. For example, increasing the salt tolerance of rice (Oryza sativa) through genetic engineering could allow weedy wild rice relatives, eg. O. rufipogon, to invade marshlands and displace native species.

Animal GMOs

The attitude of humans to domesticated vertebrate animals is culturally variable and often illogical. The anthropological debate has polarised people into opposing factions, one materialistic and the other maintaining that particular species are bonded to man and their consumption is socially and psychologically unacceptable. Research on human-animal bonding is increasing and is matched by increasing public awareness of animal welfare.

This awareness has been expressed through a strengthening of the animal welfare movement and for researchers a requirement for ethical animal experimentation. Human-animal bonding, contrasts with the human-plant bond, which is limited to the conservation of wild plant species, and is not linked to the choice of food. However, the issue of conserving wild vertebrate species is comparable to the conservation of wild plant species. Currently, gene technology has been proposed for the reconstitution of the extinct Tasmanian tiger, using preserved DNA. However, such an attempt is opposed by the conservation movement, which argues success would diminish the case for protecting existing wild species.

At the other end of a wide spectrum of animal products, the human use of xeno-transplants of organs from the unbonded pig, although showing success, is declining in acceptability because of the risk of transferring viral disease. The domestication of wild animals for soil tillage, human food and fibre accompanied the expansion of post-Neolithic agriculture. Domestication enabled control of breeding, using traditional selection and crossing to develop special purpose strains and breeds.

From the 1940s, the rate of genetic improvement for both production and quality was enhanced by study of parentoffspring and sibling relationships for enumerating the heritability of attributes and their genetic correlations. This enabled development of quantitative breeding plans to complement traditional visual selection. By the 1970s, genetic progress was slowing and the technology of genetic engineering is being evaluated. As is the case for plants, the traditional and the new genetics are likely to fuse to restore the declining rate of breeding progress.

The NRC provides principles of risk analysis as applied to GMO release. Based on current knowledge of population genetics, domestic species and ecosystems it is possible to classify GMOs into categories of high to low probabilities of spread into the environment. The risks of possible harms can then be estimated from the probability of spread. The risk is thus the product of two probabilities, the probability of exposure and the probability of harm.

Hence, the sequence for risk analysis is to:

— identify the potential harms

— identify the hazards which may produce these harms

— determine the likelihood of exposure

— quantify the likelihood of harm from that exposure multiply the probabilities of the above four parameters to prioritise risk.

There is a need to continually update this procedure as new knowledge appears. Harm can generally be regarded as species or community perturbations that result in ecological community instability. Generally, the risk assessment principles lead to the consideration of the following variables:

— the effect of the new gene(s) on the fitness of the GMO in the ecosystem

— the ability of the GMO to escape

— the stability of the existing community.

Once a new gene is introduced into a community via a GMO, natural selection for fitness will determine the ultimate fate of that gene, assuming that the initial population is large enough to cope with the initial perturbations associated with introduction.

Fitness is determined by six components: juvenile and adult viability, age at sexual maturity, female fecundity, male fertility and mating success. If an organism is fitter than its relatives in the receiving population then the GMO will eventually replace or at least establish within that community. If it is less fit, then it will be removed from the population. If fitness is similar then the GMO will persist along with the non-GMO population.

GMOs are often adapted to a wider range of environmental conditions, such as having the ability to obtain phosphorus from phytic acid. This would allow these organisms to access phosphorus from sources from which it would normally be unavailable. A new gene which increases fitness (adaptation) increases the likelihood of establishment. Many domesticated farm animals have been modified to enhance production traits. Generally, most of these traits have been selected for by conventional means, and this selection in most cases reduces fitness of the animals to the natural environment, largely because of physiologic imbalances with available nutrients in the natural environment. GMOs selected for these traits may well be even less fit to these environments.

The dangers of animals novel to the Australian environment escaping and establishing feral populations is well known (eg. rabbits, pigs etc.). Any modification to species that readily produces feral populations that makes them even more fit to the natural environment could be very detrimental. However, most modifications to these species for production purposes decreases their fitness, and hence are not a problem. A few exceptions may be the pig and mouse modified for phytase-based digestion.

The modification of insects to reduce the risk of disease transmission, such as the mosquito modified to prevent carriage of the malaria parasite, has potential ecological risks. This mosquito then is advantaged (due to no parasites) and an increase in mosquito numbers may result, leading in turn to increases in other mosquitoborne diseases. Modification of insects for biocontrol of other insects is also an area being investigated at present. However, no current guidelines exist for environmental risk assessment for release of GM arthropods.

Genetically modified aquatic organisms for aquaculture generally have a high escape potential and, if fitter for the Australian environment, would be an area of major concern. Crustaceans such as crayfish have the ability to travel overland and many molluscs have planktonic larvae that are very difficult to confine. There is a need for research into determining the ecological hazards of transgenic aquatic organism escape into the Australian ecosystem.

Across the international spectrum there has been a range of GM research directed at enhancing animal production. Myostatin acts as an inhibitor to muscle growth in animals. In beef cattle, this inhibition increases both the number and the diameter of muscle filaments. There are traditional breeds of cattle that naturally carry genes for double-muscling, producing carcasses with a high ratio of muscle to fat. A double-muscled sire is an unusual looking animal and would probably appear as freakish to the public. There is strong likelihood of public resistance to such a product as there has been for battery hens and is likely to be for Israel's production of featherless chickens.

Wool growth depends substantially on supply of the amino-acid cysteine. Sheep rely on bacteria in their alimentary tract to supply cysteine. Bacterial genes for cysteine production have been successfully inserted into mice. The mice have been able to produce their own cysteine, enabling increase in growth of their fur. The next step is to transfer this model to sheep. Sheep carrying a modified growth hormone have been developed and are undergoing trials, under secure conditions, at Armidale and Perth. The sheep are healthy and more productive than their 'unmodified cousins'.

GM aquatic species: As catches from marine fisheries decline, the development of aquaculture is assuming a greater importance. The Pacific oyster is suited to aquaculture and is regarded as a gourmet food. However, it is an invasive species and is becoming established in colonies of native rock oysters along the east coast. A GM Pacific oyster with gene insertion to curtail adult breeding, if they should escape from aqua-farms. Success would assist the balance between the demands of production, food quality and conservation of native species. Gene technology is also being used to improve the resistance of prawns and oysters to viral diseases.

9

Remediation of Metals-Contaminated Soils and Groundwater

Surface water and groundwater may be contaminated with metals from wastewater discharges or by direct contact with metals-contaminated soils, sludges, mining wastes, and debris. Metal-bearing solids at contaminated sites can originate from a wide variety of sources in the form of airborne emissions, process solid wastes, sludges or spills. The contaminant sources influence the heterogeneity of contaminated sites on a macroscopic and microscopic scale. Variations in contaminant concentration and matrix influence the risks associated with metal contamination and treatment options.

Airborne sources of metals include stack or duct emissions of air, gas, or vapor streams, and fugitive emissions such as dust from storage areas or waste piles. Metals from airborne sources are generally released as particulates contained in the gas stream. Some metals such as arsenic, cadmium, and lead can also volatilize during high-temperature processing. These metals will convert to oxides and condense as fine particulates unless a reducing atmosphere is maintained.

Stack emissions can be distributed over a wide area by natural air currents until dry and/or wet precipitation mechanisms remove them from the gas stream. Fugitive emissions are often distributed over a much smaller area because emissions are made near the ground. In general, contaminant concentrations are lower in fugitive emissions compared to stack emissions. The type and concentration of metals emitted from both types of sources will depend on site-specific conditions.

Process Solid Wastes

Process solid wastes can result from a variety of industrial processes. These metal-bearing solid wastes are disposed above ground in waste piles or below ground or under cover in landfills. Examples of process solid wastes include slags, fly ash, mold sands, abrasive wastes, ion exchange resins, spent catalysts, spent activated carbon, and refractory bricks. The composition of the process waste influences the density, porosity, and leach resistance of the waste and must be considered in evaluating the contaminated matrix.

Because waste piles are above ground, they are exposed to weathering which can disperse the waste pile to the surrounding soil, water and air and can result in generation of leachate which infiltrates into the subsurface environment. The ability of landfills to contain process solid wastes varies due to the range of available landfill designs. Uncontained landfills can release contaminants into infiltrating surface water or groundwater or via wind and surface erosion.

Sludges

The composition of sludges depends on the original waste stream and the process from which it was derived. Sludges resulting from a uniform wastestream, such as wastewater treatment sludges, are typically more homogeneous and have more uniform matrix characteristics. Sludge pits, on the other hand, often contain a mixture of wastes that have been aged and weathered, causing a variety of reactions to occur. Sludge pits often require some form of pretreatment before wastes can be treated or recycled.

Soils

Soil consists of a mixture of weathered minerals and varying amounts of organic matter. Soils can be contaminated as a result of spills or direct contact with contaminated waste streams such as airborne emissions, process solid wastes, sludges, or leachate from waste materials. The solubility of metals in soil is influenced by the chemistry of the soil and groundwater. Factors such as pH, Eh, ion exchange capacity, and complexation/chelation with organic matter directly affect metal solubility.

Contamination of Ground-Water

Groundwater can be contaminated with metals directly by infiltration of leachate from land disposal of solid wastes, liquid sewage or sewage sludge, leachate from mine tailings and other mining wastes, deep-well disposal of liquid wastes, seepage from industrial waste lagoons, or from other spills and leaks from industrial metal processing facilities (e.g., steel plants, plating shops, etc.). A variety of reactions may occur which influence the speciation and mobility of metal contaminants including acid/base, precipitation/dissolution, oxidation/ reduction, sorption or ion exchange.

Precipitation, sorption, and ion exchange reactions can retard the movement of metals in groundwater. The rate and extent of these reactions will depend on factors such as pH, Eh, complexation with other dissolved constituents, sorption and ion exchange capacity of the geological materials, and organic matter content. Ground-water flow characteristics also influence the transport of metal contaminants.

Sludges, soils, and solid wastes are multiphase materials which may contain metals in the solid, gaseous, or liquid phases. This complicates analysis and interpretation of reported results. For example, the most common method for determining the concentration of metals contaminants in soil is via total elemental analysis. The level of metal contamination determined by this method is expressed as mg metal/kg soil. This analysis does not specify requirements for the moisture content of the soil and may therefore include soil water. This measurement may also be reported on a dry soil basis.

The level of contamination may also be reported as leachable metals as determined by leach tests, such as the toxicity characteristic leaching procedure, or TCLP test or the synthetic precipitation leaching procedure, or SPLP test. These procedures measure the concentration of metals in leachate from soil contacted with an acetic acid solution (TCLP) or a dilute solution of sulfuric and nitric acid (SPLP). In this case, metal contamination is expressed in mg/L of the leachable metal. Other types of leaching tests have been proposed, including sequential extraction procedures and extraction of acid volatile sulfide.

Sequential procedures contact the solid with a series of extractant solutions that are designed to dissolve different fractions of the associated metal. These tests may provide insight into the different forms of metal contamination present. Contaminant concentrations can be measured directly in metals-contaminated water. These concentrations are most commonly expressed as total dissolved metals in mass concentrations (mg/L or :g/L) or in molar concentrations (moles/L). In dilute solutions, a mg/L is equivalent to one part per million (ppm), and a :g/L is equivalent to one part per billion (ppb).

Ground-water samples are usually filtered with a 0.45 :m filter prior to analysis for metals, though this is not always required and has recently been prohibited by many states and some U.S. EPA programs that require analysis of total metals. Interest in measurement of total metal concentrations (dissolved and particulate-associated metals) usually derives from concern about possible transport of metals adsorbed on mobile colloidal particles. Research indicates that significant colloid-facilitated transport of metals can occur only under a fairly specialized set of conditions, but the conservative approach in monitoring system design is to try to capture any mobile colloids present.

The problem with sampling groundwater without filtration is that particles from the well material, well slime coatings, or well pack may be sampled, and any subsequent analysis will not accurately reflect ground-water composition. To avoid such artifacts, but still permit sampling that can capture any mobile colloids present in the groundwater, monitoring wells are purged before sampling to remove the casing water and obtain representative ground-water samples. Low-flow purging and sampling techniques have been developed to minimize sample disturbances that may affect analysis.

Mobility of Metals in Ground-water

The fate and transport of a metal in soil and groundwater depends significantly on the chemical form and speciation of the metal. The mobility of metals in ground-water systems is hindered by reactions that cause metals to adsorb or precipitate, or chemistry that tends to keep metals associated with the solid phase and prevent them from dissolving. These mechanisms can retard the movement of metals and also provide a long-term source of metal contaminants.

While the various metals undergo similar reactions in a number of aspects, the extent and nature of these reactions varies under particular conditions. The chemical form and speciation of some of the more important metals found at contaminated sites are described below. The influence of chemical form on fate and mobility of these compounds is also explained.

Lead

The primary industrial sources of lead (Pb) contamination include metal smelting and processing, secondary metals production, lead battery manufacturing, pigment and chemical manufacturing, and lead-contaminated wastes. Widespread contamination due to the former use of lead in gasoline is also of concern. Lead released to groundwater, surface water and land is usually in the form of elemental lead, lead oxides and hydroxides, and leadmetal oxyanion complexes. Lead occurs most commonly with an oxidation state of 0 or +II. Pb(II) is the more common and reactive form of lead and forms mononuclear and polynuclear oxides and hydroxides.

Under most conditions Pb^{2+} and lead-hydroxy complexes are the most stable forms of lead. Low solubility compounds are formed by complexation with inorganic Cl^- , CO_3^{2-}, SO_4^{2-}, PO_4^{3-}) and organic ligands (humic and fulvic acids, EDTA, amino acids). Lead carbonate solids form above pH 6 and PbS is the most stable solid when high sulfide concentrations are present under reducing conditions. Most lead that is released to the environment is retained in the soil. The primary processes influencing the fate of lead in soil include adsorption, ion exchange, precipitation, and complexation with sorbed organic matter.

These processes limit the amount of lead that can be transported into the surface water or groundwater. The relatively volatile organolead compound tetramethyl lead may form in anaerobic sediments as a result of alkyllation by microorganisms. The amount of dissolved lead in surface water and groundwater depends on pH and the concentration of dissolved salts and the types of mineral surfaces present. In surface water and ground-water systems, a significant fraction of lead is undissolved and occurs as precipitates ($PbCO_3$, Pb_2O, $Pb(OH)_2$, $PbSO_4$), sorbed ions or surface coatings on minerals, or as suspended organic matter.

Chromium

Chromium(Cr) is one of the less common elements and does not occur naturally in elemental form, but only in compounds. Chromium is mined as a primary ore product in the form of the mineral chromite, $FeCr_2O_4$. Major sources of Cr contamination include releases from electroplating processes and the disposal of chromium containing wastes. Cr(VI) is the form of chromium commonly found at contaminated sites. Chromium can also occur in the +III oxidation state, depending on pH and redox conditions. Cr (VI) is the dominant form of chromium in shallow aquifers where aerobic conditions exist.

Cr(VI) can be reduced to Cr(III) by soil organic matter, S^{2-} and Fe^{2+} ions under anaerobic conditions often encountered in deeper groundwater. Major Cr(VI) species include chromate (CrO_4^{2-}) and dichromate ($Cr_2O_7^{2-}$) which precipitate readily in the presence of metal cations (especially Ba^{2+}, Pb^{2+}, and Ag^+). Chromate and dichromate also adsorb on soil surfaces, especially iron and aluminum oxides. Cr(III) is the dominant form of chromium at low pH (<4). Cr^{3+} forms solution complexes with NH_3, OH^-, Cl^-, F^-, CN^-, SO_4^{2-}, and soluble organic ligands.

Cr(VI) is the more toxic form of chromium and is also more mobile. Cr(III) mobility is decreased by adsorption to clays and oxide minerals below pH 5 and low solubility above pH 5 due to the formation of $Cr(OH)_3(s)$. Chromium mobility depends on sorption characteristics of the soil, including clay content, iron oxide content and the amount of organic matter present. Chromium can be transported by surface runoff to surface waters in its soluble or precipitated form. Soluble and unadsorbed chromium complexes can leach from soil into groundwater. The leachability of Cr(VI) increases as soil pH increases. Most of chromium released into natural waters is particle associated, however, and is ultimately deposited into the sediment.

Arsenic

Arsenic (As) is a semimetallic element that occurs in a wide variety of minerals, mainly as As_2O_3, and can be recovered from processing of ores containing mostly copper, lead, zinc, silver and gold. It is also present in ashes from coal combustion. Arsenic exhibits fairly complex chemistry and can be present in several oxidation states (-III, 0, III, V). In aerobic environments, As(V) is dominant, usually in the form of arsenate (AsO_4^{3-}) in various protonation states: H_3AsO_4, $H_2AsO_4^-$, $HAsO_4^{2-}$, AsO_4^{3-}.

Arsenate, and other anionic forms of arsenic behave as chelates and can precipitate when metal cations are present. Metal arsenate complexes are stable only under certain conditions. As(V) can also coprecipitate with or adsorb onto iron oxyhydroxides under acidic and moderately reducing conditions. Coprecipitates are immobile under these conditions but arsenic mobility increases as pH increases. Under reducing conditions As(III) dominates, existing as arsenite (AsO_3^{3-}) and its protonated forms: H_3AsO_3, $H2AsO^{3-}$, $HAsO_3^{2-}$.

Arsenite can adsorb or coprecipitate with metal sulfides and has a high affinity for other sulfur compounds. Elemental arsenic and arsine, AsH_3, may be present under extreme reducing conditions. Biotransformation (via methylation) of arsenic creates methylated derivatives of arsine, such as dimethyl arsine $HAs(CH_3)_2$ and trimethylarsine $As(CH_3)_3$ which are highly volatile. Since arsenic is often present in anionic form, it does not form complexes with simple anions such as Cl^- and SO_4^{2-}. Arsenic speciation also includes organometallic forms such as methylarsinic acid $(CH_3)AsO_2H_2$ and dimethylarsinic acid $(CH_3)_2AsO_2H$.

Many arsenic compounds sorb strongly to soils and are therefore transported only over short distances in groundwater and surface water. Sorption and coprecipitation with hydrous iron oxides are the most important removal mechanisms under most environmental conditions. Arsenates can be leached easily if the amount of reactive metal in the soil is low. As(V) can also be mobilized under reducing conditions that encourage the formation of As(III), under alkaline and saline conditions, in the presence of other ions that compete for sorption sites, and in the presence of organic compounds that form complexes with arsenic.

Zinc

Zinc (Zn) does not occur naturally in elemental form. It is usually extracted from mineral ores to form zinc oxide (ZnO). The primary industrial use for Zinc is as a corrosion-resistant coating for iron or steel. Zinc usually occurs in the +II oxidation state and forms complexes with a number of anions, amino acids and organic acids. Zn may precipitate as $Zn(OH)_2(s)$, $ZnCO_3(s)$, ZnS(s), or $Zn(CN)_2(s)$. Zinc is one of the most mobile heavy metals in surface waters and groundwater because it is present as soluble compounds at neutral and acidic pH values.

At higher pH values, zinc can form carbonate and hydroxide complexes which control zinc solubility. Zinc readily precipitates under reducing conditions and in highly polluted systems when it is present at very high concentrations, and may coprecipitate with hydrous oxides of iron or manganese. Sorption to sediments or suspended solids, including hydrous iron and manganese oxides, clay minerals, and organic matter, is the primary fate of zinc in aquatic environments. Sorption of zinc increases as pH increases and salinity decreases.

Cadmium

Cadmium (Cd) occurs naturally in the form of CdS or $CdCO_3$. Cadmium is recovered as a by-product from the mining of sulfide ores of lead, zinc and copper. Sources of cadmium contamination include plating operations and the disposal of cadmium-containing wastes. The form of cadmium encountered depends on solution and soil chemistry as well as treatment of the waste prior to disposal The most common forms of cadmium include Cd^{2+}, cadmium-cyanide complexes, or $Cd(OH)_2$ solid sludge. Hydroxide ($Cd(OH)_2$) and carbonate ($CdCO_3$) solids dominate at high pH whereas Cd^{2+} and aqueous sulfate species are the dominant forms of cadmium at lower pH (<8).

Under reducing conditions when sulfur is present, the stable solid CdS(s) is formed. Cadmium will also precipitate in the presence of phosphate, arsenate, chromate and other anions, although solubility will vary with pH and other chemical factors. Cadmium is relatively mobile in surface water and ground-water systems and exists primarily as hydrated ions or as complexes with humic acids and other organic ligands. Under acidic conditions, cadmium may also form complexes with chloride and sulfate.

Cadmium is removed from natural waters by precipitation and sorption to mineral surfaces, especially oxide minerals, at higher pH values (>pH 6). Removal by these mechanisms increases as pH increases. Sorption is also influenced by the cation exchange capacity (CEC) of clays, carbonate minerals, and organic matter present in soils and sediments. Under reducing conditions, precipitation as CdS controls the mobility of cadmium.

Copper

Copper (Cu) is mined as a primary ore product from copper sulfide and oxide ores. Mining activities are the major source of copper contamination in groundwater and surface waters. Other sources of copper include algicides, chromated copper arsenate (CCA) pressuretreated lumber, and copper pipes. Solution and soil chemistry strongly influence the speciation of copper in ground-water systems. In aerobic, sufficiently alkaline systems, $CuCO_3$ is the dominant soluble copper species. The cupric ion, Cu^{2+}, and hydroxide complexes, $CuOH^+$ and $Cu(OH)_2$, are also commonly present. Copper forms strong solution complexes with humic acids.

The affinity of Cu for humates increases as pH increases and ionic strength decreases. In anaerobic environments, when sulfur is present CuS(s) will form. Copper mobility is decreased by sorption to mineral surfaces. Cu2+ sorbs strongly to mineral surfaces over a wide range of pH values. The cupric ion (Cu^{2+}) is the most toxic species of copper. Copper toxicity has also been demonstrated for $CuOH^+$ and $Cu_2(OH)_2^{2+}$.

Mercury

The primary source of mercury is the sulfide ore cinnabar. Mercury (Hg) is usually recovered as a by-product of ore processing. Release of mercury from coal combustion is a major source of mercury contamination. Releases from manometers at pressure measuring stations along gas/oil pipelines also contribute to mercury contamination. After release to the environment, mercury usually exists in mercuric (Hg^{2+}), mercurous (Hg_2^{2+}), elemental (Hgo), or alkyllated form (methyl/ethyl mercury). The redox potential and pH of the system determine the stable forms of mercury that will be present.

Mercurous and mercuric mercury are more stable under oxidizing conditions. When mildly reducing conditions exist, organic or inorganic mercury may be reduced to elemental mercury, which may then be converted to alkyllated forms by biotic or abiotic processes.

Mercury is most toxic in its alkyllated forms which are soluble in water and volatile in air. Hg(II) forms strong complexes with a variety of both inorganic and organic ligands, making it very soluble in oxidized aquatic systems.

Sorption to soils, sediments, and humic materials is an important mechanism for removal of mercury from solution. Sorption is pH-dependent and increases as pH increases. Mercury may also be removed from solution by coprecipitation with sulfides. Under anaerobic conditions, both organic and inorganic forms of mercury may be converted to alkyllated forms by microbial activity, such as by sulfur-reducing bacteria. Elemental mercury may also be formed under anaerobic conditions by demethylation of methyl mercury, or by reduction of Hg(II). Acidic conditions ($pH<4$) also favor the formation of methyl mercury, whereas higher pH values favor precipitation of HgS(s).

Effects of Soil Properties on Mobility

Chemical and physical properties of the contaminated matrix influence the mobility of metals in soils and groundwater. Contamination exists in three forms in the soil matrix: solubilized contaminants in the soil moisture, adsorbed contaminants on soil surfaces, and contaminants fixed chemically as solid compounds. The chemical and physical properties of the soil will influence the form of the metal contaminant, its mobility, and the technology selected for remediation.

The presence of inorganic anions (carbonate, phosphate, sulfide) in the soil water can influence the soil's ability to fix metals chemically. These anions can form relatively insoluble complexes with metal ions and cause metals to desorb and/or precipitate in their presence. Soil pH values generally range between 4.0 and 8.5 with buffering by Al at low pH and by $CaCO_3$ at high pH. Metal cations are most mobile under acidic conditions while anions tend to sorb to oxide minerals in this pH range. At high pH, cations precipitate or adsorb to mineral surfaces and metal anions are mobilized.

The presence of hydrous metal oxides of Fe, Al, Mn can strongly influence metal concentrations because these minerals can remove cations and anions from solution by ion exchange, specific adsorption and surface precipitation. Cation exchange capacity (CEC) refers to the concentration of readily exchangeable cations on a mineral surface and is often used to indicate the affinity of soils for uptake of cations such as metals. Anion exchange capacity (AEC) indicates the affinity of soils for uptake of anions, and is usually significantly lower than the CEC of the soil. In addition to hydrous oxides, clays are also important ion exchange materials for metals. The presence of natural organic matter (NOM) has been shown to influence the sorption of metal ions to mineral surfaces. NOM has been observed to enhance sorption of Cu^{2+} at low pH, and suppress Cu2+ sorption at high pH.

Organic matter, particularly humic materials, can complex metals and affect their removal from solution. Humic materials contain carboxylic and phenolic functional groups that can complex with metal ions.

Particle size distribution can influence the level of metal contamination in a soil. Fine particles (<100 :m) are more reactive and have a higher surface area than coarser material. As a result, the fine fraction of a soil often contains the majority of contamination. The distribution of particle sizes with which a metal contaminant is associated can determine the effectiveness of a number of metal remediation technologies, e.g., soil washing. Soil moisture influences the chemistry of contaminated soil.

The amount of dissolved minerals, pH and redox potential of the soil water depends on the soil moisture content. Soil structure describes the size, shape, arrangement and degree of development of soils into structural units. Soil structure can influence contaminant mobility by limiting the degree of contact between groundwater and contaminants.

Technical Remediations for Metal Contamination

The physical and chemical form of the metal contaminant in soil or water strongly influences the selection of the appropriate remediation treatment approach. Information about the physical characteristics of the site and the type and level of contamination at the site must be obtained to enable accurate assessment of site contamination and remedial alternatives. The importance of adequate, well-planned site characterization to selection of an appropriate cost-effective remediation approach has been discussed many times but cannot be overemphasized.

The contamination in the groundwater and soil should be characterized to establish the type, amount, and distribution of contaminants across different media. Once the site has been characterized, the desired level of each contaminant in soil and groundwater must be determined. This is done by comparison of observed contaminant concentrations with soil and ground-water quality standards for a particular regulatory domain, or by performance of a site-specific risk assessment. Remediation goals for metals may be set as desired concentrations in groundwater, as total metal concentration in soil, as leachable metal in soil, or as some combination of these.

Several technologies exist for the remediation of metals-contaminated soil and water. These technologies are contained within five categories of general approaches to remediation: isolation, immobilization, toxicity reduction, physical separation and extraction. These are the same general approaches used for many types of contaminants in the subsurface. As is usually the case, combinations of one or more of these approaches are often used for more cost-effective treatment of a contaminated site.

Isolation

Isolation technologies attempt to prevent the transport of contaminants by containing them within a designated area. These technologies can be used to prevent further contamination of groundwater when other treatment options are not physically or economically feasible for a site. Contaminated sites may also be isolated temporarily in order to limit transport during site assessment and site remediation.

Capping

Capping systems are used to provide an impermeable barrier to surface water infiltration to contaminated soil for prevention of further release of contaminants to the surrounding surface water or groundwater. Secondary objectives include controlling gas and odor emissions, improving aesthetics, and providing a stable surface over a contaminated site. Capping also eliminates risks associated with dermal contact and/or incidental ingestion of surface soils, but if this is the primary goal for the site and surface water infiltration is not a concern, a less expensive permeable cover may be preferred. Capping provides a range of design options that includes simple single-layer caps and more complex multilayer systems.

Design selection depends on site characteristics, remedial objectives and risk factors associated with the site. A variety of materials are available for use in capping systems and choice of materials is site specific because local soils are often incorporated into parts of the cap. Synthetic membranes such as high-density polyethylene are also available for incorporation into capping systems.

Surface water controls, such as ditches and dikes are usually included to help control drainage from the cap. Multilayered capping systems may also include a hard cover and/or a layer of topsoil to separate the underlying layers from the ground surface. Revegetation is promoted in order to reinforce the topsoil, to reduce soil erosion and runoff velocity, and to help remove water from the soil by evapotranspiration.

Subsurface Barriers

Subsurface barriers may be used to isolate contaminated soil and water by controlling the movement of groundwater at a contaminated site. These barriers are designed to reduce the movement of contaminated groundwater from the site, or to restrict the flow of uncontaminated groundwater through the contaminated site. Vertical barriers are commonly used to restrict the lateral flow of groundwater. For effective isolation of the contaminated matrix, the barrier should extend and key into a continuous, low-permeability layer, such as clay or competent bedrock, below the contaminated area. If an impermeable layer is not available, a ground-water extraction system must be used to prevent transport of contaminants under the barrier.

Vertical barriers may be installed upstream, downstream, or completely surrounding the site and are often implemented in conjunction with a capping system to control surface water infiltration. The use of circumferential barriers can prevent the escape of contamination from the site by using an infiltration barrier and collection system to create a hydraulic gradient in the inward direction. Vertical barriers are often limited to depths achievable with backhoe excavation technology for trenches, i.e., to about 30 feet.

Slurry walls are usually constructed in a vertical trench excavated under a slurry that is designed to prevent collapse and to form a filter cake on the walls of the trench to prevent the loss of fluids to the surrounding soil. A vibrating beam method (Slurry Systems, Inc.) is also available in which the beam penetrates the ground and slurry materials are injected into the soil (with assistance from a high pressure/low volume jet if needed). Two options exist for the slurry composition. The soil-bentonite (SB) slurry wall is the most common type, and comprises a bentonite-water slurry that is mixed with a soil engineered to harden upon addition to the slurry.

The trench can also be excavated under a portland cement-bentonite-water slurry that is left to harden and form a cement-bentonite (CB) slurry wall. Available technologies for installation of slurry walls allow installation to depths up to 125 feet. Slurry walls are the most common type of vertical barrier due to their low relative cost. The use of slurry walls can be limited by the topography, geology, and type of contamination at the site. For example, an SB slurry will flow unless the site and confining layer are nearly level. Also, some contaminants, such as concentrated organics and strong acids/bases, can degrade SB materials and prevent the application of SB slurry walls at some sites.

Other available vertical barriers include grout curtains and sheet piles. Grout curtains are constructed by drilling a borehole and injecting a fluid into the surrounding soil that is designed to solidify and reduce water flow through the contaminated region. The fluid is pressure-injected in rows of staggered boreholes that are designed to overlap once the fluid has permeated into the surrounding soil. Common materials used to construct grout curtains include cement, clays, alkali-silicate, and organic polymers.

Clays are the most widely used grouting materials due to their low cost. This technique is more expensive than slurry walls and its use is therefore usually limited to sealing voids in existing rock. Sheet piles usually comprise steel pilings that are driven into the formation to create a wall to contain the groundwater. Sheet piles are seldom used at contaminated sites due to concerns about wall integrity. This method is generally limited to isolation of shallow contamination (40-50 ft) distributed over a relatively small area, or used in conjunction with a soil-bentonite slurry when site conditions prevent the use of conventional slurry walls. Technologies for the construction of horizontal barriers are under investigation.

Horizontal barriers would enable control of the downward migration of contaminants by lining the site without requiring excavation of the contaminated matrix. The technologies under investigation include grout injection by vertical boring and horizontal drilling. The vertical boring method is similar to the construction of grout curtains except that the grout is injected at a fixed elevation over a tightly spaced grid of vertical boreholes to create an impermeable horizontal layer. Problems with this method include soil compaction by the large drill rigs situated over the contaminated area.

Also, the vertical boreholes would provide access to the deeper layers and may therefore increase vertical migration of contaminants. Horizontal drilling involves the use of directional drilling techniques to create the horizontal grout layer. Horizontal barriers may also be used in conjunction with vertical barriers at sites where a natural aquitard is not present. In this case, the vertical barrier could key into the horizontal barrier to prevent the transport of contaminants under the vertical barrier.

Immobilization

Immobilization technologies are designed to reduce the mobility of contaminants by changing the physical or leaching characteristics of the contaminated matrix. Mobility is usually decreased by physically restricting contact between the contaminant and the surrounding groundwater, or by chemically altering the contaminant to make it more stable with respect to dissolution in groundwater. The aqueous and solid phase chemistry of metals is conducive to immobilization by these techniques.

A variety of methods are available for immobilization of metal contaminants, including those that use chemical reagents and/or thermal treatment to physically bind the contaminated soil or sludge. Most immobilization technologies can be performed ex situ or in situ. In situ processes are preferred due to the lower labor and energy requirements, but implementation in situ will depend on specific site conditions.

Solidification and Stabilization

Solidification and stabilization (S/S) immobilization technologies are the most commonly selected treatment options for metals-contaminated sites. Solidification involves the formation of a solidified matrix that physically binds the contaminated material. Stabilization, also referred to as fixation, usually utilizes a chemical reaction to convert the waste to a less mobile form. The general approach for solidification/stabilization treatment processes involves mixing or injecting treatment agents to the contaminated soils.

Inorganic binders, such as cement, fly ash, or blast furnace slag, and organic binders such as bitumen are used to form a crystalline, glassy or polymeric framework around the waste. The dominant mechanism by which metals are immobilized is by precipitation of hydroxides within the solid matrix. S/S technologies are not useful for some forms of metal contamination, such as species that exist as anions (e.g., Cr(VI), arsenic) or metals that don't have low-solubility hydroxides (e.g., mercury). S/S may not be applicable at sites containing wastes that include organic forms of contamination, especially if volatile organics are present.

Mixing and heating associated with binder hydration may release organic vapors. Pretreatment, such as air stripping or incineration, may be used to remove the organics and prepare the waste for metal stabilization/ solidification. The application of S/S technologies will also be affected by the chemical composition of the contaminated matrix, the amount of water present, and the ambient temperature. These factors can interfere with the solidification/stabilization process by inhibiting bonding of the waste to the binding material, retarding the setting of the mixtures, decreasing the stability of the matrix, or reducing the strength of the solidified area.

Cement-based binders and stabilizers are common materials used for implementation of S/S technologies. Portland cement, a mixture of Ca-silicates, aluminates, aluminoferrites, and sulfates is an important cement-based material. Pozzolanic materials which consist of small spherical particles formed by coal combustion (such as fly ash) and in lime and cement kilns, are also commonly used for S/S. Pozzolans exhibit cement-like properties, especially if the silica content is high. Portland cement and pozzolans can be used alone or together to obtain optimal properties for a particular site.

Organic binders may also be used to treat metals through polymer microencapsulation. This process uses organic materials such as bitumen, polyethylene, paraffins, waxes and other polyolefins as thermoplastic or thermosetting resins. For polymer encapsulation, the organic materials are heated and mixed with the contaminated matrix at elevated temperatures (120° to 200°C). The organic materials polymerize, agglomerate the waste and the waste matrix is encapsulated. Organics are volatilized and collected and the treated material is extruded for disposal or possible reuse (e.g., as paving material).

The contaminated material may require pretreatment to separate rocks and debris and dry the feed material. Polymer encapsulation requires more energy and more complex equipment than cement-based S/S operations. Bitumen (asphalt) is the cheapest and most common thermoplastic binder. S/S is achieved by mixing the contaminated material with appropriate amounts of binder/

stabilizer and water. The mixture sets and cures to form a solidified matrix and contain the waste. The cure time and pour characteristics of the mixture and the final properties of the hardened cement depend upon the composition (amount of cement, pozzolan, water) of the binder/stabilizer.

Ex situ S/S can be easily applied to excavated soils because methods are available to provide the vigorous mixing needed to combine the binder/stabilizer with the contaminated material. Pretreatment of the waste may be necessary to screen and crush large rocks and debris. Mixing can be performed via in-drum, in-plant or area mixing processes. In-drum mixing may be preferred for treatment of small volumes of waste or for toxic wastes. Inplant processes utilize rotary drum mixers for batch processes or pug mill mixers for continuous treatment.

Larger volumes of waste may be excavated and moved to a contained area for area mixing. This process involves layering the contaminated material with the stabilizer/binder, and subsequent mixing with a backhoe or similar equipment. Mobile and fixed treatment plants are available for ex situ S/S treatment. Smaller pilot-scale plants can treat up to 100 tons of contaminated soil per day, while larger portable plants typically process 500 to over 1000 tons per day. S/S techniques are available to provide mixing of the binder/stabilizer with the contaminated soil in situ. In situ S/S is less labor and energy intensive than ex situ process that require excavation, transport and disposal of the treated material.

In situ S/S is also preferred if volatile or semi volatile organics are present because excavation would expose these contaminants to the air. However the presence of bedrock, large boulders, cohesive soils, oily sands and clays may preclude the application of in situ S/S at some sites. It is also more difficult to provide uniform and complete mixing through in situ processes. Mixing of the binder and contaminated matrix may be achieved using in-place mixing, vertical auger mixing or injection grouting. In-place mixing is similar to ex situ area mixing except that the soil is not excavated prior to treatment.

The in situ process is useful for treating surface or shallow contamination and involves spreading and mixing the binders with the waste using conventional excavation equipment such as draglines, backhoes or clamshell buckets. Vertical auger mixing uses a system of augers to inject and mix the binding reagents with the waste. Larger (6-12 ft diameter) augers are used for shallow (10-40 ft) drilling and can treat 500-1000 cubic yards per day. Deep stabilization/ solidification (up to 150 ft) can be achieved by using ganged augers (up to 3 ft in diameter each) that can treat 150-400 cubic yards per day. Finally injection grouting may be performed to inject the binder containing suspended or dissolved reagents into the treatment area under pressure. The binder permeates the surrounding soil and cures in place.

Vitrification

The mobility of metal contaminants can be decreased by high-temperature treatment of the contaminated area that results in the formation of vitreous material, usually an oxide solid. During this process, the increased temperature may also volatilize and/or destroy organic contaminants or volatile metal species (such as Hg) that must be collected for treatment or disposal. Most soils can be treated by vitrification and a wide variety of inorganic and organic contaminants can be targeted. Vitrification may be performed ex situ or in situ, although in situ processes are preferred due to the lower energy requirements and cost.

Typical stages in ex situ vitrification processes may include excavation, pretreatment, mixing, feeding, melting and vitrification, off-gas collection and treatment, and forming or casting of the melted product. The energy requirement for melting is the primary factor influencing the cost of ex situ vitrification. Different sources of energy can be used for this purpose, depending on local energy costs. Process heat losses and water content of the feed should be controlled in order to minimize energy requirements.

Vitrified material with certain characteristics may be obtained by using additives such as sand, clay and/or native soil. The vitrified waste may be recycled and used as clean fill, aggregate, or other reusable materials. In situ vitrification (ISV) involves passing electric current through the soil using an array of electrodes inserted vertically into the contaminated region. Each setting of four electrodes is referred to as a melt. If the soil is too dry, it may not provide sufficient conductance and a trench containing flaked graphite and glass frit (ground glass particles) must be placed between the electrodes to provide an initial flow path for the current.

Resistance heating in the starter path melts the soil. The melt grows outward and down as the molten soil usually provides additional conductance for the current. A single melt can treat up to 1000 tons of contaminated soil to depths of 20 feet, at a typical treatment rate of 3 to 6 tons per hour. Larger areas are treated by fusing together multiple individual vitrification zones. The main requirement for in situ vitrification is the ability of the soil melt to carry current and solidify as it cools. If the alkali content (as Na_2O and K_2O) of the soil is too high (1.4 wt%) the molten soil may not provide enough conductance to carry the current.

Toxicity and Mobility Reduction

Chemical and biological processes can be used to alter the form of metal contaminants in order to decrease their toxicity and/or mobility.

Chemical reactions can be initiated that are designed to decrease the toxicity or mobility of metal contaminants. The three types of reactions that can be used for this purpose are oxidation, reduction, and neutralization reactions. Chemical oxidation changes the oxidation state of the metal atom through the loss of electrons. Commercial oxidizing agents are available for chemical treatment, including potassium permanganate, hydrogen peroxide, hypochlorite and chlorine gas.

Reduction reactions change the oxidation state of metals by adding electrons. Commercially available reduction reagents include alkali metals (Na, K), sulfur dioxide, sulfite salts, and ferrous sulfate. Changing the oxidation state of metals by oxidation or reduction can detoxify, precipitate, or solubilize the metals. Chemical neutralization is used to adjust the pH balance of extremely acidic or basic soils and/or groundwater. This procedure can be used to precipitate insoluble metal salts from contaminated water, or in preparation for chemical oxidation or reduction.

Chemical treatment can be performed ex situ or in situ. However in situ chemical agents must be carefully selected so that they do not further contaminate the treatment area. The primary problem associated with chemical treatment is the nonspecific nature of the chemical reagents. Oxidizing/reducing agents added to the matrix to treat one metal will also target other reactive metals and can make them more toxic or mobile. Also, the long-term stability of reaction products is of concern since changes in soil and water chemistry might reverse the selected reactions. Chemical treatment is often used as pretreatment for S/S and other treatment technologies.

Reduction of Cr(VI) to Cr(III) is the most common form of chemical treatment and is necessary for remediation of wastes containing Cr(VI) by precipitation or S/S. Chromium in its Cr(III) form is readily precipitated by hydroxide over a wide range of pH values. Acidification may also be used to aid in Cr(VI) reduction. Arsenic may be treatable by chemical oxidation since arsenate, As(V), is less toxic, soluble and mobile than arsenite, As(III). Bench-scale work has indicated that arsenic stabilization may be achieved by precipitation and coprecipitation with Fe(III).

Treatment Walls

Treatment walls remove contaminants from groundwater by degrading, transforming, precipitating or adsorbing the target solutes as the water flows through permeable trenches containing reactive material within the subsurface. Several methods are available for installation of permeable treatment walls, some of which employ slurry wall construction technology to create a permeable reactive curtain. The reactive zone can use physical, chemical and biological processes, or a combination of these.

The ground-water flow through the wall may be enhanced by inducing a hydraulic gradient in the direction of the treatment zone or channeling ground-water flow toward the treatment zone. Several types of treatment walls are being tried for arresting transport of metals in groundwater at contaminated sites. Trench materials being investigated include zeolite, hydroxyapatite, elemental iron, and limestone.

Biological Treatment

Biological treatment technologies are available for remediation of metals-contaminated sites. These technologies are commonly used for the remediation of organic contaminants and are beginning to be applied for metal remediation, although most applications to date have been at the bench and pilot scale. Biological treatment exploits natural biological processes that allow certain plants and microorganisms to aid in the remediation of metals. These processes occur through a variety of mechanisms, including adsorption, oxidation and reduction reactions, and methylation.

Bioaccumulation

Bioaccumulation involves the uptake of metals from contaminated media by living organisms or dead, inactive biomass. Active plants and microorganisms accumulate metals as the result of normal metabolic processes via ion exchange at the cell walls, complexation reactions at the cell walls, or intra- and extracellular precipitation and complexation reactions. Adsorption to ionic groups on the cell surface is the primary mechanism for metal adsorption by inactive biomass. Accumulation in biomass has been shown to be as effective as some ion exchange resins for metals removal from water.

Phytoremediation

Phytoremediation refers to the specific ability of plants to aid in metal remediation. Some plants have developed the ability to remove ions selectively from the soil to regulate the uptake and distribution of metals. Most metal uptake occurs in the root system, usually via absorption, where many mechanisms are available to prevent metal toxicity due to high concentration of metals in the soil and water. Potentially useful phytoremediation technologies for remediation of metals-contaminated sites include phytoextraction, phytostabilization and rhizofiltration.

Phytoextraction

Phytoextraction employs hyperaccumulating plants to remove metals from the soil by absorption into the roots and shoots of the plant. A hyperaccumulator is defined as a plant with the ability to yield 0.1% chromium, cobalt, copper or nickel or 1% zinc, manganese in the aboveground shoots on a dry weight basis. The aboveground shoots can be harvested to remove metals from the site and subsequently disposed as hazardous waste or treated for the recovery of the metals.

Phytostabilization

Phytostabilization involves the use of plants to limit the mobility and bioavailability of metals in soil. Phytostabilizers are characterized by high tolerance of metals in surrounding soils but low accumulation of metals in the plant. This technique may be used as an interim containment strategy until other remediation techniques can be developed, or as treatment at sites where other methods would not be economically feasible.

Rhizofiltration

Rhizofiltration removes metals from contaminated groundwater via absorption, concentration and precipitation by plant roots. This technique is use to treat contaminated water rather than soil and is most effective for large volumes of water with low levels of metal contamination. Terrestrial plants are more effective than aquatic plants because they develop a longer, more fibrous root system that provides a larger surface area for interaction. Wetlands construction is a form of rhizofiltration that has been demonstrated as a costeffective treatment for metals-contaminated wastewater.

Bioleaching

Bioleaching uses microorganisms to solubilize metal contaminants either by direct action of the bacteria, as a result of interactions with metabolic products, or both. Bioleaching can be used in situ or ex situ to aid the removal of metals from soils. This process is being adapted from the mining industry for use in metals remediation. The mechanisms responsible for bioleaching are not fully defined, but in the case of mercury bioreduction (to elemental mercury) is thought to be responsible for mobilization of mercury salts.

Biochemical processes

Microbially mediated oxidation and reduction reactions can be manipulated for metal remediation. Some microorganisms can oxidize/reduce metal contaminants directly while others produce chemical oxidizing/reducing agents that interact with the metals to effect a change in oxidation state. Mercury and cadmium have been observed to be oxidized through microbial processes, and arsenic and iron are readily reduced in the presence of appropriate microorganisms. The mobility of metal contaminants is influenced by their oxidation state. Redox reactions can therefore be used to increase or decrease metal mobility.

Methylation involves attaching methyl groups to inorganic forms of metal ions to form organometallic compounds. Methylation reactions can be microbially mediated. Organometallic compounds are more volatile than inorganic metals and this process can be used to remove metals through volatilization and subsequent removal from the gas stream. However, organometallics are also more toxic and mobile than other metal forms and may potentially contaminate surrounding surface waters and groundwater.

Physical Separation Techniques

Physical separation is an ex situ process that attempts to separate the contaminated material from the rest of the soil matrix by exploiting certain characteristics of the metal and soil. Physical separation techniques are available that operate based on particle size, particle density, surface and magnetic properties of the contaminated soil. These techniques are most effective when the metal is either in the form of discrete particles in the soil or if the metal is sorbed to soil particles that occur in a particular size fraction of the soil. Physical separation is often used as a form of pretreatment in order to reduce the amount of material requiring subsequent treatment.

Several techniques are available for physical separation of contaminated soils including screening, classification, gravity concentration, magnetic separation and froth flotation. Screening separates soils according to particle size by passing the matrix through a sieve with particular size openings. Smaller particles pass through the sieve and leave larger particles behind, however, the separation is not always complete. Screening may be performed as a stationary process or with motion using a wet or dry process stream.

Classification involves separation of particles based upon the velocity with which they fall through water (hydroclassification) or air (air classification). Hydroclassification is more common for soil separation and may be performed using a non-mechanical, mechanical or a hydraulic classifier. Gravity concentration relies on gravity and one or more other forces (centrifugal force, velocity gradients, etc.) that may be applied to separate particles on the basis of density differences.

Gravity concentration may be achieved through the use of a hydrocyclone, jig, spiral concentrator, or shaking table. Froth flotation uses air flotation columns or cells to remove particles from water. In this process, air is sparged from the bottom of a tank or column that contains a slurry of the contaminated material. Some metals and minerals attach to the air bubbles due to particular surface properties, such as hydrophobicity. Froth flotation can be used to remove metals that attach to air bubbles, or to remove other minerals while the metal remains in the slurry.

Magnetic separation subjects particles to a strong magnetic field using electromagnets or magnetic filters and relies on differences in magnetic properties of minerals for separation. Low intensity wet magnetic separators are the most common magnetic separation devices. This process can recover a wide variety of minerals and is particularly successful for separating ferrous from nonferrous minerals.

Extraction

Metals-contaminated sites can be remediated using techniques designed to extract the contaminated fraction from the rest of the soil, either in situ or ex situ. Metal extraction can achieved by contacting the contaminated soil with a solution containing extracting agents (soil washing and in situ soil flushing) or by electrokinetic processes. The contaminated fraction of soil and/or process water is separated from the remaining soil and disposed or treated.

Soil Washing

Soil washing can be used to remove metals from the soil by chemical or physical treatment methods in aqueous suspension. Soil washing is an ex situ process that requires soil excavation prior to treatment. Chemical treatment involves addition of extraction agents that react with the contaminant and leach it from the soil. The liquid containing the contaminants is separated from the soil resulting in a clean solid phase. Physical treatment is achieved by particle size separation technologies adapted from mineral processing to concentrate the contaminant in a particular size fraction.

Fine particles (<63 :m) often contain the majority of contaminated material because they bind contaminants strongly due to their large and reactive surface area. Many current soil washing approaches attempt to separate the fine fraction from the remainder of the soil in order to reduce the amount of material for subsequent treatment or disposal. Particle size separation techniques may not be successful if fine particle, e.g., metal oxide, coatings are present on particles in larger size fractions.

Preliminary screening

After excavation, the soil undergoes preliminary screening and preparation in order to separate large rocks and debris from the contaminated matrix. Residual fines may be adhered to the surface of large rocks and are often washed off prior to return of the large rocks to the site.

Secondary screening

Most soil washing processes employ secondary screening to segregate the particles into different size fractions, usually between 5 mm and 60 mm. Most secondary screening processes involve making an aqueous slurry of the soil stream and wet screening/sieving of the slurry. The particles in this size range are considered less contaminated than the finer fraction and may be returned to the site as clean soil after separation from the water.

Chemical treatment

Chemical treatment may be used to solubilize contaminants from the most contaminated fraction of the soil. Chemical treatment is performed in an aqueous slurry of the contaminated material to which an extracting agent is added. The extraction is performed in a mixing vessel or in combination with the physical treatment stage. The type of extractant used will depend on the contaminants present and the characteristics of the soil matrix. Many processes manipulate the acid/base chemistry of the slurry to leach contaminants from the soil.

However, if a very low pH is required concerns about dissolution of the soil matrix may arise. Chelating agents (e.g., EDTA) selectively bind with some metals and may be used to solubilize contaminants from the soil matrix. Oxidizing and reducing agents (e.g., hydrogen peroxide, sodium borohydride) provide yet another option to aid in solubilization of metals since chemical oxidation/ reduction can convert metals to more soluble forms. Finally, surfactants may be used in extraction of metals from soil.

Physical treatment

Physical treatment is used to separate the contaminated fraction, usually the fine materials, from the rest of the soil matrix. Physical separation may be performed alone or in conjunction with chemical treatment, as in most soil washing processes. The most common method for physical separation in soil washing uses rotary attrition scrubbers to isolate the contaminated particles. The rotation of the slurry causes contact between large particles, resulting in attrition of the larger particles which releases the contaminant and contaminated fines to the slurry. The contaminant remains suspended in solution or sorbs to the reactive fine particles.

Vibration units are also available to perform similar separations. Hydrocyclones are the most common method used to separate fines from the clean soil. Other options are available for fine particle separation, including mechanical classifiers, gravity classifiers, spiral concentrators, and magnetic separators. Froth flotation can be used to combine physical and chemical treatment processes into one step. For this method, extracting agent is added to the soil before it enters the froth flotation cell. The slurry is leached in the tanks to remove the contaminant and the fines (<50 :m) are then separated from coarse particles in the flotation unit.

After the contaminated fine particles are separated from the clean coarse particles, both fractions are dewatered. The fine fraction is usually dewatered using a belt filter or filter press and disposed of in a landfill. Larger particles are rinsed to remove residual extracting solution and contaminant and dewatered using belt and filter presses. This fraction is considered clean and can be returned to the site.

The contaminated water from rinsing and dewatering steps is treated by manipulating the solution chemistry to separate the contaminant from the extractant if possible. Contaminants can then be removed from solution, most commonly by precipitation or sedimentation, and are dewatered before disposal with the contaminated fines. The extracting agent and process water can be recycled for reuse.

Pyrometallurgical Technologies

Pyrometallurgical technologies use elevated temperature extraction and processing for removal of metals from contaminated soils. Soils are treated in a high-temperature furnace to remove volatile metals from the solid phase. Subsequent treatment steps may include metal recovery or immobilization. Pyrometallurgical treatment requires a uniform feed material for efficient heat transfer between the gas and solid phases and minimization of particulates in the off-gas. This process is usually preceded by physical treatment to provide optimum particle size. Pyrometallurgical processes usually produce a metal-bearing waste slag, but the metals can also be recovered for reuse.

In Situ Soil Flushing

In situ soil flushing is used to mobilize metals by leaching contaminants from soils so that they can be extracted without excavating the contaminated materials. An aqueous extracting solution is injected into or sprayed onto the contaminated area to mobilize the contaminants usually by solubilization. The extractant can be applied by surface flooding, sprinklers, leach fields, vertical or horizontal injection wells, basin infiltration systems or trench infiltration systems. After being contacted with the contaminated material the extractant solution is collected using pump-and-treat methods for disposal or treatment and reuse.

Similar extracting agents are used for in situ soil flushing and soil washing, including acids/bases, chelating agents, oxidizing/reducing agents and surfactants/cosolvents. Also, water can be used alone to remove water-soluble contaminants such as hexavalent chromium. The applicability of in situ soil flushing technologies to contaminated sites will depend largely on site-specific properties, such as hydraulic conductivity, that influence the ability to contact the extractant with contaminants and to effectively recover the flushing solution with collection wells.

Electrokinetic Treatment

Electrokinetic remediation technologies apply a low density current to contaminated soil in order to mobilize contaminants in the form of charged species. The current is applied by inserting electrodes into the subsurface and relying on the natural conductivity of the soil (due to water and salts) to effect movement of water, ions and particulates through the soil. Water and/or chemical solutions can also be added to enhance the recovery of metals by this process. Positively charged metal ions migrate to the negatively charged electrode, while metal anions migrate to the positively charged electrode.

Electrokinetic treatment concentrates contaminants in the solution around the electrodes. The contaminants are removed from this solution by a variety of processes, including electroplating at the electrodes, precipitation/coprecipitation at the electrodes, complexation with ion exchange resins, or by pumping the water from the subsurface and treating it to recover the extracted metals. Electrokinetic treatment is most applicable to saturated soils with low ground-water flow rates and moderate to low permeability. The efficiency of metal removal by this process will be influenced by the type and concentration of contaminant, the type of soil, soil structure, and interfacial chemistry of the soil.

Commercially Available Technologies

Ex situ treatment technologies are examined only for soils. The full range of contaminated water treatment technologies is available for ex situ treatment of groundwater. For the most part, the technologies reported are those whose performance has been verified by the U.S. EPA under the Superfund Innovative Technology Evaluation (SITE) program which evaluates emerging and demonstrated technologies.

Superfund Innovative Technology Evaluation (SITE)

The 1986 Superfund Amendments and Reauthorization Act recognized a need for an "alternative or innovative technology research and demonstration program." In response, the U.S. EPA established the Superfund Innovative Technology Evaluation (SITE) Program to encourage the development and implementation of innovative treatment technologies for remediation of hazardous waste sites and for monitoring and measurement. Innovative technologies are field-tested in the SITE Demonstration Program and engineering and cost data are collected to assess the performance of the technology.

The demonstration stage also attempts to evaluate the applicability of the technology to different types of wastes and waste matrices, the need for pre- and post-processing of the waste stream, and potential operating problems. The Resource Conservation and Recovery Act (RCRA) provides for determination of a Best Demonstrated Available Technology (BDAT) for treatment of hazardous wastes.

BDATs have been established based upon critical analysis of performance data collected for treatment of various industry-generated wastes. BDAT status is given only to proven, commercially available technologies.

Different BDATs and treatment standards are usually given for nonwastewater and wastewater forms of contamination. The applicability of a BDAT to metals-contaminated soil and water at a Superfund site must be evaluated on a site specific basis. The establishment of a BDAT does not prevent the use of other available technologies for treatment of these wastes.

Containment

Containment technologies are widely used to control the transport of hazardous materials and prevent the spread of contamination. Containment is the preferred remedial method for sites having low levels of wastes with low toxicity and low mobility, or wastes that have been pretreated to obtain these characteristics. Containment may also be used as a temporary measure to reduce the mobility of wastes that pose a high risk until a permanent remedy is selected and implemented.

Advantages to containment technologies include relatively simple and rapid implementation often at lower cost than alternatives that require excavation; ability to treat large areas and volumes of waste; and the potential for successful containment as the final action at the site. Uncertainty regarding long-term effectiveness and the need for long-term inspection because untreated contaminants remain onsite are among the disadvantages of containment technologies. Also, future use of the site may be limited if containment technologies are used.

Capping systems have been selected for a number of sites with low levels of metal contamination. Monitoring wells and/or infiltration monitoring systems are often used to help assess the performance of capping systems. Slurry walls have also been used for containment of metals-contaminated sites. The performance of vertical containment barriers also must be monitored. Performance can be influenced by geography, topography, and geology. The presence of certain compounds can also influence the long-term integrity of some cement-based vertical barriers by chemically attacking the soil-bentonite blends.

Material availability can affect the application of slurry walls and other containment technologies. There are no established BDAT's for containment technologies since they are not considered to be treatment technologies. Ongoing SITE demonstrations for remediation of metals by containment technologies include a high clay grouting procedure (Morrison Knudsen Corporation) and frozen soil barriers.

Ex Situ Remediation

The majority of the technologies that have been demonstrated for metals remediation to date are ex situ technologies. Ex situ remediation technologies demonstrated include solidification/stabilization, soil washing, vitrification and pyrometallurgic separation.

Immobilization technologies

Immobilization technologies, especially solidification/stabilization, are the most common methods selected for remediation of metal contamination, accounting for nearly 30 percent of all soil treatment technologies at Superfund sites. S/S techniques have been widely used to manage metal wastes at hazardous waste sites and to treat residues from other treatment processes. Benefits associated with immobilization treatments include their broad application to a wide variety of metals and also to wastes that contain mixtures of metals and organics.

Solidification/stabilization technologies using cement-based and pozzolan binders are available commercially and have been applied at several sites for a wide variety of metals, including chromium, lead, arsenic, mercury, and cadmium. Remediation has been completed for a number of these sites and S/S has been selected or initiated for several others. SITE demonstrations have been performed or are underway for various ex situ stabilization/solidification technologies.

Ex situ solidification/stabilization techniques have been determined to be the BDAT for a range of waste types, including cadmium nonwastewaters (other than Cd-containing batteries), chromium nonwastewaters (after reduction to Cr(III)), lead nonwastewaters, wastes with low (<260mg/kg) concentrations of elemental mercury, and plating and steelmaking wastes. S/S can also be used to treat arsenic wastes even though vitrification was selected as the BDAT for arsenic-containing nonwastewaters.

Soil washing

Soil washing technologies are applicable to a range of soils containing a variety of metal contaminants. Soil washing is most easily implemented when a single metal contaminant occurs in a particular insoluble fraction of the soil which can be separated by particle size classification. Soils with a minimum content of finer material (<20% of particles with diameters <2 mm) are easier to process. Soil washing has been used for remediation of metals-contaminated sites in Europe and has been selected and/or implemented at several U.S. Superfund sites.

Remediation at the Twin Cities Army Ammunition Plant (TCAAP) in New Brighton, Minnesota is one of the two completed soil washing projects. The COGNIS TERRAMET soil washing procedure used at this site employed a combination of particle sizing, gravity separation, and acid-leaching apparatus that was designed to remove lead, mercury, cadmium, chromium and copper from the soil. Preliminary studies have shown that the primary target metal at this site, lead, could be reduced from over 86000 mg/kg to less than 100 mg/kg, well below the target cleanup level of 300 mg/kg. Acid leaching soil washing procedures have been designated as the BDAT for mercury-contaminated soils. Several SITE demonstrations have been performed for soil washing of metalscontaminated soils.

Vitrification

Vitrification is most applicable to sites containing low-volatility metals with high glass solubilities, and therefore appears to be well-suited for treatment of lead, chromium, arsenic, zinc, cadmium and copper wastes. The ability of a melt to retain these and other metals depends on the metal solubility in the soil at the site, and silica content of the soil. The metal concentration can be adjusted by adding soil or another source of silica to improve site characteristics for vitrification. The ability to control volatile emissions also influences the applicability of vitrification technologies.

Mercury's high volatility and low glass solubility makes it unsuitable for vitrification, but treatment by vitrification may be allowed at sites containing very low mercury concentrations. Ex situ vitrification may not be applicable for soils with greater than 25% moisture content due to excess fuel consumption, or at sites where size reduction and classification are not feasible. Several ex situ vitrification technologies are under development. SITE program demonstrations have been completed for two of these processes and a third demonstration is underway. Ex-situ vitrification has also been demonstrated for treatment of cesium-contaminated tank wastes from the Oak Ridge Reservation.

Pyrometallurgical separation

Mercury has a relatively high vapor pressure and is easily converted to its metallic form at elevated temperature, making it easily treated by pyrometallurgic methods. Pyrometallurgic treatment of lead, arsenic, cadmium and chromium may require pretreatment by reducing agents or fluxing agents to facilitate melting. Nonvolatile metals such as chromium can be tapped from the furnace as molten metal.

Pyrometallurgical treatment is usually performed offsite because few mobile treatment units are available. This technology is most applicable to large volumes of highly-contaminated soils (metal concentrations >5%-20%, especially when metal recovery is expected. Low metal concentrations can be processed, especially for mercury since it is easy to volatilize and recover. A number of pyrometallurgical process technologies are currently available for treatment of metals-contaminated soils.

Pyrometallurgical treatment is a BDAT for cadmium-containing batteries, lead nonwastewaters, mercury wastes, mercury from wastewater treatment sludge, lead acid batteries, and zinc nonwastewaters. SITE demonstrations have been completed for thermal desorption (RUST Remedial Services, Inc.) and flame reactor (Horsehead Resource Development Company, Inc.) pyrometallurgical technologies.

In Situ Remediation

In situ remediation technologies offer the potential for significant cost savings over ex situ technologies because in situ techniques are usually associated with lower labor and energy requirements for implementation.

Solidification

In situ S/S treatment appears to have been applied less

frequently than ex situ techniques mostly due to concerns about uniformity of treatment and long-term reliability. These limitations are being reduced, however, through advances in chemical reagent delivery systems for large-diameter auger drilling devices. While in situ S/S technologies are well developed due to roots in construction techniques, data on the performance of in situ S/S are limited. Based upon preliminary data, in situ S/ S appears likely to be an effective treatment option.

In situ S/S typically will be most beneficial for sites with contamination at depths less than 8-10 feet and for larger volumes of waste because ex situ may prove to be cheaper for small volumes and shallow contamination due to high costs associated with mobilization and demobilization for in situ technologies. Deep soil mixing technology is also available for treating contaminated soils at greater depth but is more expensive than shallow soil mixing. The cost of in situ technologies is also affected by implementation concerns such as a level, stable base that is required for augering, and the presence of large rocks that can make large-diameter augering impossible. The use of dry reagents in soils with high moisture content is a well established method in Europe that is gaining interest in the U.S. and may expand the applicability of in situ S/S techniques.

Vitrification

In situ vitrification (ISV) technologies are currently offered commercially in the U.S. by a single vendor, Geosafe Corporation. The first full-scale application of ISV was demonstrated at the Parsons Chemical/ETM Enterprises Superfund site in Grand Ledge, Michigan under the EPA SITE program. The Geosafe ISV process was used for treatment of soils and sediments contaminated with pesticides, metals (As, Cr, Hg, Pb), and dioxins. This treatment system required the use of eight melts which were each completed over a time frame of 10 to 20 days. This system also included an air emissions control system to treat volatilized contaminants, including mercury.

While ISV is not recommended for remediation of mercury, this method can be used in conjunction with emissions control systems when Hg is present in mixed metal/organic wastes. This treatment was successful, meeting TCLP limits for all of the metals in the treated waste. ISV has also been used successfully at two sites contaminated with organics (PCB, dioxin, pentachlorophenol, pesticides, herbicides), further demonstrating the applicability of this technology. Based upon observations from these limited applications, it appears that ISV may not be appropriate for sites with high levels of organics (> 10 % organics by weight) due to contamination of the off-gas, or inorganics (> 25 % metals by weight, or > 20 % by volume) due to concerns about exceeding glass solubility limits.

Soil flushing

In Situ soil flushing has been selected for treatment at several Superfund sites contaminated with metals. In situ soil flushing is the technology in design or the predesign stage at least five other sites. This technology has been applied for a limited number of projects, mostly containing organic forms of contamination, and limited information is available on the application of this technology to metals-contaminated sites. The United Chrome Products Superfund site in Corvallis, Oregon is currently being remediated using in situ soil flushing technologies.

The soil and groundwater at this site are heavily contaminated with chromium, with chromium levels in the soil as high as 60000 mg/kg and levels in the groundwater reaching up to 19000 mg/kg. The general approach to remediation of this site has been removal of the more soluble, mobile and toxic form of chromium, Cr(VI), by flushing the contaminated region with water to solubilize Cr(VI), with subsequent extraction of the chromium-containing water for treatment. Remediation at this site began in 1985 and has combined a variety of technologies to aid remediation by in situ soil flushing. The technologies used have included infiltration basins and trenches to flush contaminated soils, a 23-well ground-water extraction network to remove contaminated groundwater and recharge water, on-site treatment of wastewater, and off-site disposal of contaminated soil and debris.

This full-scale application of in situ soil flushing with water as the flushing solution appears to be successful for removal of Cr(VI) from coarse soils of relatively high hydraulic conductivity. The in situ soil flushing procedure used at this site leaches contaminants from the unsaturated and saturated zones, and provides for recharge of the groundwater to the extraction wells. This cleanup operation has removed significant amounts of chromium from the soil and groundwater and the ground-water pumping strategy has achieved hydraulic containment of the plume. Cr(VI) levels in water retrieved by the extraction wells decreased from more than 5000 mg/L to approximately 50 mg/L during the first two and one half years of operation.

Average chromium concentrations in the plume decreased from 1923 mg/L to 207 mg/L after flushing the first one and one half pore volumes (approximately 2.6 million gallons for one pore volume). These rapid removal rates are expected to continue for the first few pore volumes of treatment until Cr(VI) removal begins to tail off to the asymptotic level. Tailing results from slow desorption from soil particles, dissolution of solid phase contaminants, and release of contaminants from the fine pores in the soil matrix. Tailing is commonly observed in in situ soil flushing applications and usually represents the practical limit for remediation via pump and treat methods.

Electrokinetic extraction

The success of various electrokinetic remediation technologies has been illustrated for removal of metals from soils via bench and pilot scale experiments. Currently, several of these technologies are being implemented in comprehensive demonstration studies to further the use of electrokinetic techniques at contaminated sites. Electrokinetic remediation of metals-contaminated sites has been demonstrated in situ at many sites in Europe using processes developed by Geokinetics International, Inc. (GII). The success of electrokinetic remediation appears to vary depending on the metals present, and can remove up to 90% of the initial contamination.

Biological treatment

Phytoremediation technologies are largely in the developmental stage and many are being field tested at a variety of sites in the U.S. and in Europe. Because full-scale applications of phytoremediation technologies are just being initiated, limited cost and performance data are available. Some techniques under development have shown potential for use at metalscontaminated sites. Phytostabilization and phytoextraction methods are being developed by Phytotech, Inc. and field tests for patented phytoextraction techniques are being performed.

Some grasses have been made commercially available for phytostabilization of metals (lead, copper, zinc). Nickel has been removed from plating wastes by bacteria and other organisms are being genetically engineered to remove metals such as cadmium, cobalt, copper and mercury. Bioreduction has been demonstrated (for Hg) at the bench scale but has not been tested at pilot scale. A process has been developed for chromium reduction by H_2S produced by sulfate-reducing bacteria and reduction of Cr(VI) by direct metabolism is being investigated by several organizations.

Treatment by wetlands has been studied under the U.S. EPA's SITE program. Full-scale demonstration of a constructed wetland is planned for the Burleigh Tunnel site, part of the Clear Creek/Central City Superfund site in Colorado. Bioleaching is currently used to recover copper and uranium ores by heap or in situ leaching and is under development for a wide range of metals including cadmium, chromium, lead, mercury and nickel.

Microorganisms have been tested for chemical reduction and removal of mercury salts from wastewater. Phytoremediation technologies will likely be limited to use in shallow soils with relatively low levels of metal contamination. Based upon estimates of biomass productivity and metal content of soils, the annual removal rate of metals by phytoremediation would be limited to between 2.5 to 100 mg/kg of soil contaminants.

10

Hazardous Wastes

All waste generators, except households, are required by law to determine if any of their waste is hazardous waste. The wastes are hazardous if they are either included on specific lists or exhibit certain hazardous characteristics. These wastes have been determined to be a threat to human health or the environment. In addition, hazardous waste includes mixtures resulting from combining a listed waste with another waste unless that mixture is exempted from the hazardous waste regulations.

Hazardous wastes have specific numbers assigned to the different constituents or processes that generated the waste. Depending on the waste, you may have some waste to which several waste numbers apply. Regulations allow businesses the option of handling some waste as a "universal waste" instead of managing them as a hazardous waste.

Listed Waste

Listed waste includes waste materials listed by name or generation source on the federal and Michigan lists of hazardous waste. You will need to know the chemical names, and in some instances you will also need to know the chemical concentrations for the type of waste you produce, and/or the process you use to determine if it is a listed waste. It will be necessary to review the regulations for the complete description of these listed wastes. These wastes have an EPA or Michigan hazardous waste number, which begins or ends with the letters as indicated below:

Waste from listed nonspecific sources: also referred to as the "F" list. Vehicle service businesses primarily have F001-F005 wastes that are spent solvents that were used for their degreasing and cleaning properties. These solvents must contain the constituents included in the regulatory descriptions, and if found in a mixture or blend, then it must also meet the applicable concentration level before it was used. For example, if a solvent product (that contained methyl ethyl ketone (MEK) and other listed solvents which resulted in a blend that was over 10% by volume of the product) was used to clean out a paint gun, then the paint booth filter would be an F005 waste because the solvent was used for its cleaning properties.

However, if a paint product that contained MEK was used for its intended purpose, the paint booth filter waste would not be a "F" listed waste because the solvent is not being used as a degreasing or cleaning agent. Depending on the vehicle business activities, occasionally other "F" code wastes may be generated. A few "F" wastes, those that have a (H) included in their hazard code, are also considered acutely hazardous.

Waste from listed specific sources: also referred to as the "K" list. Vehicle service related operations normally do not have K code wastes.

Discarded commercial chemical products and spill residues: also referred to as the "P" and "U" lists. Chemicals included on the "P" list are considered acutely hazardous. "U" wastes include toxic chemicals and chemicals that also display other characteristics such as ignitability. Some of these discarded chemicals used by vehicle operations would only be a U or P code waste if the listed chemical is the sole active ingredient in the product, or the waste was in a pure or technical grade solution. For example, if you use technical grade toluene for cleaning, this would be a U220 waste if the product was discarded before being used. It would be an F005 waste if it was used before being discarded. Vehicle related businesses normally have P or U code wastes only if disposing of unused or off-spec chemicals or are cleaning up a spill.

Characteristics of Waste

Characteristic waste exhibiting any of five characteristics identified in the Michigan and federal regulations is also defined as a hazardous waste. These wastes have an EPA or Michigan hazardous waste number which begins or ends with a "D" or "S". The five characteristics are:

Ignitable: Starts burning easily; liquids with a flash point below 140 degrees Fahrenheit, solid that spontaneously ignites, or oxidizing material. Examples include gasoline, cleaning fluids, naphtha, methyl isobutyl ketone, mineral spirits, sludges containing petroleum, solvents, solvent based paints, solvent-soaked rags, and ignitable compressed gas like hydrogen. These wastes have a hazardous waste number of D001.

Corrosive: Liquids that dissolve metals or aqueous wastes with a pH less than or equal to 2.0 or greater than or equal to 12.5. Examples include caustics like alkaline cleaners and battery acid. These wastes have a hazardous waste number of D002.

Reactive: Undergoes rapid or violent chemical reaction and necessitates special handling requirements. Examples include organic peroxides, cyanides, sulfides, and explosives. These wastes have a hazardous waste number of D003.

Toxic: Poisonous to humans and other living organisms. These wastes become regulated as a hazardous waste when their constituents meet or exceed a certain concentration level. These are sometimes called toxicity characteristic leaching procedure (TCLP) wastes. Hazardous waste numbers are assigned to specific toxic chemicals and include D004 through D043. Fluorescent lamps, dry cell batteries, various metal-bearing solutions, solvents, mercury switches, lead tire weights, some pesticides, and other organic chemicals are some materials that may be toxic wastes. An example of a D009 hazardous waste would include mercury from electric lamps or switches that had a TCLP test concentration result of 0.2 milligrams per liter (mg/l) or more. MEK (methyl ethyl ketone) would have a waste number of D035 if its TCLP concentration was 200 mg/l or more and could also be a "F" waste if it met any of the listed waste descriptions.

Severely toxic: These Michigan hazardous wastes contain 1.0 ppm or more of a severely toxic material. These materials are regulated at quantities of one kilogram, which is just over two pounds or more. The hazardous waste numbers include 001S through 007S. It is unlikely vehicle service businesses would have severely toxic wastes.

Recycling

Some hazardous waste that is being recycled in specific situations is excluded. This includes solvents, oils and filters, lead acid batteries, scrap metal, textiles, and spent chlorofluorocarbon refrigerants. Generally to meet the recycling exclusions, the recovered material must meet the following conditions:

— It is used as an ingredient to make a product, without first being reclaimed;

— It is used as an effective substitute for a commercial chemical product; or ·It is returned to the original process from which it was generated, without first being reclaimed.

Wastewater containing hazardous waste discharged to publicly owned treatment plants (POTW) are exempt from the hazardous waste regulations at the point of discharge into the sewer system. However, any hazardous waste generation, treatment, or storage prior to that discharge is subject to the hazardous waste regulations. This exemption does not apply to hazardous waste that is transported by truck or rail to a POTW. An exemption from the mixture rule exists if very small amounts, or deminimus amounts, of listed hazardous waste are discharged to a facility's wastewater treatment plant with large volumes of non hazardous wastewater. This exemption probably won't apply to most vehicle service industries.

Empty containers, liners, and residue from empty containers are not subject to the hazardous waste requirements if the following conditions are met:

1. The containers that held non acute hazardous waste have had as much material removed as possible by practices commonly used to remove that material, such as pouring, pumping, and aspirating, and the amount of residue in containers holding the hazardous waste is any of the following:

   — One inch or less; OR

   — No more than three percent by weight of the total capacity for containers 110 gallons or less in size; OR

   — No more than 0.3 percent by weight of the total capacity for containers over 110 gallons.

2. The containers that held acutely or severely toxic hazardous waste (e.g., waste identified on the "P" or "S" lists, and some "F" wastes) have been triple-rinsed using a material capable of removing the product, or by another proven cleaning method, or the inner lining that prevented contact of the chemical with the container has been removed from the container.

3. Compressed gas cylinders have been emptied to the point when the pressure in the containers approaches atmospheric pressure.

A waste sample that is sent to a laboratory to determine if a waste stream is a hazardous waste is exempt from most of the hazardous waste regulations if it meets certain conditions. It is necessary to only send the smallest amount needed for the test (typically this is less than one gallon) to the laboratory and the laboratory must send back any remaining sample to the generator. If the sample is determined to be a hazardous waste, then this exemption no longer applies to the sample after it is returned to the generator and it is no longer needed for waste characterization purposes.

When shipping the sample, it is necessary to meet any US Postal Service or USDOT labeling and shipping requirements. If these agencies' regulations do not apply to the sample, then the sample must be packed so that it does not leak, spill, or vaporize. In addition, the following information must accompany the shipment:

— Sample collector's name, mailing address, and telephone number

— Laboratory's name, mailing address, and telephone number

— Date of shipment

— Quantity of the sample

— Description of the sample

Universal Waste

Many businesses are not aware that Michigan adopted the universal waste rule in October 1996 and many companies are not properly handling these wastes. This rule allows businesses the choice of handling specific wastes as a universal waste or they can continue to manage them as a hazardous waste. The following may be handled as universal waste:

— Electric lamps including fluorescent, high intensity discharge, sodium vapor, mercury vapor, neon, and incandescent lamps, and cathode ray tubes from computers, televisions, or other equipment;

— Batteries, including lead acid and dry cell types;

— Devices containing only elemental mercury, such as thermostats, switches, thermometers; and other devices;

— Pesticides, including certain suspended, canceled, or unused pesticides.

There are two (2) levels of universal waste handlers-a Small Quantity Handler and Large Quantity Handler. A Small Quantity Handler accumulates less than 5,000 kilograms (11,000 pounds) of universal waste at any time. A Large Quantity Handler accumulates 5,000 kilograms (11,000 pounds) or more of universal waste at any time.

The following identifies some advantages of handling these wastes as universal waste:

— If a company chooses to manage these waste streams as a universal waste, they do not include this quantity in determining their generator status. For some companies this may allow them to reduce their generator status level. For example, a Large Quantity Generator that manages part of its hazardous waste stream as a universal waste, may be able to become a Small Quantity Generator.

— Facilities can accumulate universal wastes up to one year. This is a longer storage time than allowed for Small Quantity and Large Quantity Generators' hazardous waste.

— There is less labeling required for storage of universal waste than what is required for hazardous waste. The labeling is specific to the waste.

— A hazardous waste manifest and the use of permitted and registered hazardous waste transporter is not necessary to ship the waste.

Hazardous waste may be generated in many areas of our business from the shop floor to office space. A few activities that may generate hazardous wastes are:

— Use shop rags to apply or wipe off solvents or other hazardous chemicals

— Use parts washers or other solvents included on the "F" lists in the hazardous waste regulations. Spent solvents commonly found include those containing toluene and MEK that may have a hazardous waste number of F005; methylene chloride which is often found in carburetor cleaners that may have a waste number of F001, and xylene and acetone that may have a waste number of F003.

— Reduce inventory and discard unused chemical product; such as any of the commercial chemical products on the "P" and "U" lists in the state or federal regulations. Examples include 1,1,1-Trichloroethane that has a hazardous waste number of U228, xylene with a waste number of U239, and toluene with a waste number of U220.

— Use paint products including paint thinners, enamel reducers, epoxys, primers, enamels and other solvent based paints or that contain heavy metals

— Use rust removing agents like naval jelly, strong acids or bases

— Mix listed wastes with another waste. The entire mixture (both the listed and other wastes) would now be considered a hazardous waste unless it meets an exemption.

— Change light bulbs

— Upgrade your office computer equipment that may contain lead in the cathode ray tubes, mercury switches, batteries

— Replace lead acid batteries from vehicles or discard portable service tools that contains dry cell batteries

Following are some commonly overlooked wastes from a vehicle service business:

— Spent fluorescent tubes and other lighting fixtures (toxic for mercury);

— Disposable rags containing free liquids with a flash point of less than 140 degrees Fahrenheit or used with a listed solvent (ignitability via spontaneous combustion, used with "F" listed solvents);

— Spent activated carbon media, included in some air filters (contains "F" listed solvents);

— Used solvents with low flashpoint (toxic, ignitability);

— Drain or sump sludge (contains toxic metals or "F" solvents, ignitability due to gasoline);

— Painting materials and waste (contains "F" solvents, metals, ignitability);

— Aerosol cans that are not empty (contains "F" solvents, ignitability);

— Solvent based adhesives (contains "F" solvents, ignitability);

— Batteries-lead acid and dry cell (toxic for lead and mercury, corrosive);

— Used water-based or synthetic lubricating fluids containing high concentrations of heavy metals (toxic metals of concern include lead, chromium, cadmium, and barium); and

— Painted automotive glass (toxic for lead).

After identifying all the hazardous wastes generated at your business, then determine if any regulatory exemptions or exclusions apply to those wastes.

If you know or suspect that you have not fully evaluated your waste or if the wastes are complex, get help. Waste evaluation services can often be obtained from a testing laboratory, consultant, intended treatment or disposal facility, or provided by your transporter. Sometimes it will be necessary to test samples of your waste to determine if it is hazardous waste. When you need to do this, you should hire a reputable firm to provide these services and obtain a written contract. The contract should clearly identify what specific services that company will provide.

It is a good idea to check with the treatment, storage, and disposal facility (TSDF) where you intend to send your waste before hiring a testing laboratory. They might require specific laboratory tests and only accept data from specific laboratories. If that is the case, then ask them for a listing of these tests and what is the purpose of the tests, along with the approved testing methods, and the acceptable laboratories. This step will prevent you from spending money on laboratory tests that are not necessary or do not meet the TSDF's requirements. Be aware that it is not necessary to identify every chemical component of the waste in order to meet regulations and ensure adequate treatment or disposal.

The paint filter test is a method used to determine the presence of free liquids in a representative sample of waste. A predetermined amount of material is placed in a paint filter. If any portion of the material passes through and drops from the filter within the 5-minute test period, it contains free liquids. If the waste is not regulated under the hazardous waste regulations, then it would be regulated under Part 121 as a liquid industrial waste.

A Toxicity Characteristic Leaching Procedure (TCLP) is used to determine if a waste would have toxicity characteristics in amounts that meet or exceed regulatory limits and be regulated as hazardous waste. The TCLP was designed to predict whether a waste is likely to leach chemicals into groundwater. It simulates the conditions a waste might encounter in a typical municipal landfill. Be aware that it is not necessary to identify every chemical component of the waste in order to meet the hazardous waste regulations and ensure adequate treatment or disposal. It may not be necessary to run a TCLP for every constituent included on the "D" list if you are familiar with your process. For example, you may only need to have a TCLP done for metals and volatiles if you know that the other constituents would not be present in the waste.

In other situations, you may only need to know if a liquid waste is ignitable and could request a flash point test; or you may need to know if it is corrosive and have a pH test done. Special tests might be required if you have drums or containers of mixed or unidentified old waste. You may be able to minimize the laboratory testing costs by providing information about your waste streams and operations that were previously collected during your waste survey.

Hazardous Waste Generators

There are three categories of hazardous waste generators. The company's hazardous waste status is based on the total quantity of all the hazardous waste being generated and accumulated at your site over a specific time period that does not meet an exclusion or exemption.

Vehicle service businesses are likely to be required to obtain either a Michigan Liquid Industrial Waste Generator

Identification Number (number begins with MIG), or an EPA Identification Number (number usually begins with MIR or MID). These numbers are used to track liquid industrial waste and hazardous waste. The type of identification number you need is based on the type of waste generated, and your generator status or universal waste handler level. If you generate hazardous waste, you need to obtain the appropriate number for your business as identified below:

If you are a Conditionally Exempt Small Quantity Generator, you are not required by law to apply for an EPA Identification Number for shipping your hazardous waste. You may use either a MIG or an EPA Identification Number when shipping liquid industrial or hazardous waste. If you have an EPA Identification Number, then you MUST use that number on your manifests. Some disposal companies might require you to have an EPA Identification Number.

If you are a Small Quantity or a Large Quantity Generator of hazardous waste, or a Large Quantity Handler of universal waste, you must notify the WMD that you are generating hazardous waste or handling universal waste and obtain an EPA Identification Number. A Small Quantity Handler of universal waste is not required to get an EPA Identification Number. The EPA number will also be used when shipping liquid industrial waste. Transporters and facilities that treat, store or dispose of regulated quantities of hazardous waste and universal waste destination facilities must also have an EPA Identification Number.

Bibliography

Anderson, T.A., and Coats, J.R., eds. "Bioremediation through Rhizosphere Technology." ACS Symposium Series No. 563. American Chemical Society, Washington, D.C., 1994.

Atlas, R.M. "Bioremediation". *Chemical and Engineering News,* April, 32 42, 1995.

de Lorenzo, V. "Genetic engineering strategies for environmental applications". *Current Opinion in Biotechnology* 3, 227 231, 1992.

Economidis, I., and Palmisano, A., eds. "Proceedings: Environmental Biotechnology: EC-US Task Force on Biotechnology Research." Brussels, Belgium, 1995.

Ehrlich, H.L. "Geomicrobiology." Marcel Dekker, Inc., New York, 1990.

Francis, A.J. *Microbial dissolution and stabilization of toxic metals and radionuclides in mixed wastes.* Experientia 46, 840 851, 1990.

Frankenberger, Jr., W.T., and Benson, S.M., eds. "Selenium in the Environment." Marcel Dekker, Inc., New York, 1994.

Grimes, D.J., "Validation of reliability and predictability of environmental biotechnologies", in *Proceedings of the 1994 OECD Workshop on Bioremediation.* Tokyo, Japan.

King, R.B., Mihony, G., and Sheldon, J.K. "Practical Environmental Bioremediation." Lewis Publishers, Boca Raton, Florida, 1992.

Lynch, J.M., and Hobbie, J.E. "Micro-organisms in Action: Concepts and Applications in Microbial Ecology."

Blackwell Scientific Publications, Oxford, England, 1988.

Macaskie, L.E., "The application of biotechnology to the treatment of waste products produced from the nuclear fuel cycle: Biodegradation and bioaccumulation as a means of treating radionuclide-containing streams". *Critical Reviews in Biotechnology* 11, 41 112, 1991.

Metting, Jr., F.B., ed., "Soil Microbial Ecology: Applications in Agricultural and Environmental Management." Marcel Dekker, Inc., New York, 1992.

Miller, R.V., and Poindexter, J.S., "Strategies and Mechanisms for Field Research in Environmental Bioremediation." American Academy of Microbiology, Washington, D.C., 1994.

Mueller, J.G. "Model studies of bioremediation strategies and monitoring technologies, in "Proceedings: Environmental Biotechnology: EC-US Task Force on Biotechnology Research" (I. Economidis and A. Palmisano, eds.). Brussels, Belgium, 1995.

Sayler, G.S., "Biodegradation process analysis: Molecular application in simulations and environmental verification", *in "Proceedings: Environmental Biotechnology: EC-US Task Force on Biotechnology Research* (I. Economidis and A. Palmisano, eds.), 47 54. Brussels, Belgium, 1995.

Sayler, G.S., Fox, R., and Blackburn, J.W. *Environmental Biotechnology for Waste Treatment*, Plenum Press, New York, 1991.

Shannon, M.J.R., and Unterman, R. "Evaluating bioremediation". *Annual Review of Microbiology* 47, 715 730, 1993.

Skladany, G.J., and Metting, Jr., F.B. "Bioremediation of contaminated soil", in *Soil Microbial Ecology: Applications in Agricultural and Environmental Management* (F.B. Metting, Jr., ed.), 483 513. Marcel Dekker, Inc., New York.

Thomas, J.M., and Ward, C.H., "In situ biorestoration of organic contaminants in the subsurface". *Environmental Science and Technology* 23, 760 766, 1989.

Index